Starting with the Unit Circle
Background to Higher Analysis

Loo-keng Hua

Starting with the Unit Circle
Background to Higher Analysis

Translated by Kuniko Weltin

With 6 Illustrations

Springer-Verlag
New York Heidelberg Berlin

Loo-keng Hua
Institute of Applied Mathematics
Academia Sinica
Beijing
People's Republic of China

Kuniko Weltin
10 The Crescent
Berkeley, CA 94708
USA

AMS Classification (1980): 26-01, 20C30, 31A05, 35M05, 33A99, 51N30, 83A05

Library of Congress Cataloging in Publication Data
Hua, Loo-keng, 1911–
    Starting with the unit circle.

    Translation of: Ts'ung tan wei yüan t'an ch'i.
    1. Mathematical analysis.   1. Title.
QA300.H8613   515   81-8840
                                      AACR2

Printed in the United States of America.

9 8 7 6 5 4 3 2 1

ISBN 0-387-90589-8 Springer-Verlag New York Heidelberg Berlin
ISBN 3-540-90589-8 Springer-Verlag Berlin Heidelberg New York

Contents

CHAPTER 3
Extended Space and Spherical Geometry

CHAPTER 4
The Lorentz Group

CHAPTER 5
The Fundamental Theorem of Spherical Geometry—with a
Discussion of the Fundamental Theorem of Special Relativity

CHAPTER 6
Non-Euclidean Geometry

CHAPTER 7
Partial Differential Equations of Mixed Type

CHAPTER 8
Formal Fourier Series and Generalized Functions

Preface

It is with great pleasure that I am writing the preface for my little book, "Starting with the Unit Circle", in the office of Springer Verlag in Heidelberg. This is symbolic of the fact that I have once again joined in the mainstream of scientific exchange between East and West.

Since the establishment of the People's Republic of China, I have written "An Introduction to Number Theory" for the young people studying Number Theory: for the young people studying algebra, Prof. Wan Zhe-xian (Wan Che-hsien) and I have written "Classical Groups"; for those studying the theory of functions of several complex variables, I have written "Harmonic Analysis of Functions of Several Complex Variables in the Classical Domains", * and for university students I have written "Introduction to Higher Mathematics". The present volume had been written for those who were beginning to engage in research at the Chinese University of Science and Technology and at the Guangdong Zhongshan University. Its purpose is none other than to make the students see the crucial ideas in their simplest manifestations, so that when they go on to the more complex parts of modern mathematics, they will not be without guidance.

For example, in the first chapter when I point out that the Poisson kernel is just the Jacobian of some transformation, I am merely revealing the source of one of the main tools in my work on harmonic analysis in the classical domains.

In the second chapter we treat the expansion problem of harmonic functions. This is in fact the starting point of my research in the study of Fourier series on the unitary group by using group representation theory.

---

* English translation, American Mathematical Society, 1963.

The Riemann mapping theorem tells us that if we are given a simply connected region on a plane, then its topological transformations and its holomorphic transformations are closely related. However, as we move into higher dimensions, they get farther apart. From the topological viewpoint, in the one variable case, the Alexandroff compactification (of adding a point at infinity) of $\mathbb{C}$ gives rise to a homogeneous space. However, in the case of several complex variables and algebraic geometry, we find that a totally different phenomenon occurs, and the problem arises as to how to compactify by adding more than one point, so that the space becomes a homogeneous space under an analytic transformation group or an algebraic group. Chapters Three and Six give a brief exposition of this problem. In Chapter Five we emphasize a special example. Not only do we give examples of matrix geometry over a skew field, but we also point out its relationship with the theory of special relativity. I am grateful to Prof. Gunsen of Birmingham University for bringing to my attention the work done on this problem by Prof. Zeemann in 1964. Although our results were obtained somewhat earlier, by comparison they are not without merit.

The idea behind Chapter Seven is a continuation of that of Chapter One, except that it is now applied to partial differential equations of mixed type. The reason we deliberately avoid the method of the Beltrami operator of Riemannian geometry is to enable the readers to fully comprehend the basic principles from concrete examples. Actually, there is ample room for generalization in this direction.

Naturally, the boundary value problem of elliptic equations is, given a function defined on the boundary, whether or not a solution of the elliptic equation exists in the bounded domain. On the other hand, having a solution does not have to mean assuming the boundary value in the usual sense. We define this type of function to be a generalized function, and in Chapter Eight we discuss the simple and concrete aspect of generalized functions.

As for the words, "Starting with," in the title of the book, they are intended to indicate that further development is possible. We have carried out part of the development in this book, but it is quite evident that much more can be done.

Finally, I would like to express my heartfelt thanks to Springer-Verlag, and especially to Dr. Götze. Due to his outstanding organizational abilities, not only is my book appearing in English, but so will quite a few other Chinese books as well.

I would also like to thank Professors Gong Shen, Wu Ziqian, Lin Wei and Chen De-quan, for without their encouragement and help, this book would never have seen its day.

Hua Loo-keng
Heidelberg
Nov. 17, 1979

Translator's note. An effort has been made in this translation to remain as faithful as possible to Professor Hua's original text. Only such changes have been made as those which must accompany any translation between languages with drastically different syntax and idioms. There are also minor improvements on the format of the original, such as italicizing key terms, displaying most of the theorems, adding an index and incorporating the usual American notation for cross-references, e.g. (3.9) refers to equation (9) in §3 of the same chapter. References to items in a different chapter are explicitly stated as such.

CHAPTER 1
The Geometric Theory of Harmonic Functions

1.1 Remembrance of Things Past

Consider the transformation on the complex plane

$$w = \frac{z - a}{1 - \bar{a}z}, \qquad |a| < 1, \tag{1}$$

where $\bar{a}$ denotes the conjugate of a, and the transformation

$$w = e^{i\theta}z. \tag{2}$$

From (1) it follows that

$$1 - |w|^2 = 1 - \frac{(z - a)(\bar{z} - \bar{a})}{(1 - \bar{a}z)(1 - a\bar{z})}$$

$$= \frac{(1 - |a|^2)(1 - |z|^2)}{|1 - a\bar{z}|^2}. \tag{3}$$

Since transformation (1) transforms the unit circle $|z| = 1$ into the unit circle $|w| = 1$, the disc $|z| < 1$ is transformed into the unit disc $|w| < 1$. Transformation (2) also has this property. Furthermore, transformation (1) transforms $z = a$ into $w = 0$.

Differentiating equation (1) we obtain

$$dw = \frac{dz}{1 - \bar{a}z} + \frac{(z - a)\bar{a}\,dz}{(1 - \bar{a}z)^2} = \frac{1 - a\bar{a}}{(1 - \bar{a}z)^2}\,dz. \tag{4}$$

Dividing (4) by (3) and taking the square of the absolute value, we obtain the invariant differential form associated with transformation (1) and (2)

$$\frac{|dw|^2}{(1 - |w|^2)^2} = \frac{|dz|^2}{(1 - |z|^2)^2}. \tag{5}$$

1

Corresponding to this second order differential form, there is an invariant differential operator

$$(1 - |w|^2)^2 \frac{\partial^2 \Phi}{\partial w \, \partial \overline{w}} = (1 - |z|^2)^2 \frac{\partial^2 \Phi}{\partial z \, \partial \overline{z}}. \tag{6}$$

The operator

$$4 \frac{\partial^2 \Phi}{\partial z \, \partial \overline{z}} = \frac{\partial^2 \Phi}{\partial x^2} + \frac{\partial^2 \Phi}{\partial y^2}$$

is of course the *Laplacian*. Since (1) transforms the unit circle into the unit circle, when $z = e^{i\tau}$ ($\tau \in [0, 2\pi]$), we have $w = e^{i\psi}$ for some $\psi \in [0, 2\pi]$, i.e. we have

$$e^{i\psi} = \frac{e^{i\tau} - a}{1 - \overline{a} e^{i\tau}} = \frac{1 - a e^{-i\tau}}{1 - \overline{a} e^{i\tau}} e^{i\tau}.$$

This represents the transformation on the unit circle induced by transformation (1). Thus, (4) becomes

$$e^{i\psi} \, d\psi = \frac{1 - a\overline{a}}{(1 - \overline{a} e^{i\tau})^2} e^{i\tau} \, d\tau.$$

Dividing the two preceding equations gives

$$d\psi = \frac{1 - a\overline{a}}{|1 - \overline{a} e^{i\tau}|^2} \, d\tau. \tag{7}$$

Now let

$$a = \rho e^{i\theta}, \qquad \rho < 1,$$

and

$$P(\rho, \theta - \tau) = \frac{1 - |a|^2}{|1 - \overline{a} e^{i\tau}|^2} = \frac{1 - \rho^2}{1 - 2\rho \cos(\theta - \tau) + \rho^2}. \tag{8}$$

The latter is called the *Poisson kernel*. Thus the Poisson kernel is the Jacobian associated with transformation (1) of the unit circle into itself. The Poisson kernel has the following properties:

(i) *Positive definiteness*: for $\rho < 1$, $P(\rho, \theta - \tau) > 0$.

(ii) $\displaystyle\lim_{\rho \to 1} P(\rho, \theta - \tau) = \begin{cases} 0 & \text{for } \theta \neq \tau \\ \infty & \text{for } \theta = \tau \end{cases}$

(iii) $\displaystyle\frac{1}{2\pi} \int_0^{2\pi} P(\rho, \theta - \tau) \, d\tau = 1$

(iv) For $\rho < 1$, $P(\rho, \theta - \tau)$ satisfies the *Laplace equation* which in polar coordinates reads:

$$\rho \frac{\partial}{\partial \rho} \left(\rho \frac{\partial u}{\partial \rho} \right) + \frac{\partial^2 u}{\partial \theta^2} = 0. \tag{9}$$

(i) and (ii) are immediate. The proof of (iii) follows from (7) and

$$\frac{1}{2\pi} \int_0^{2\pi} P(\rho, \theta - \tau)\, d\tau = \frac{1}{2\pi} \int_0^{2\pi} d\psi = 1.$$

For the proof of (iv):

$$P(\rho, \theta - \tau) = 1 + \frac{\rho e^{i(\theta - \tau)}}{1 - \rho e^{i(\theta - \tau)}} + \frac{\rho e^{-i(\theta - \tau)}}{1 - \rho e^{-i(\theta - \tau)}}$$

$$= 1 + 2 \sum_{n=1}^{\infty} \rho^n \cos n(\theta - \tau);$$

since $\rho^n \cos n(\theta - \tau)$ clearly satisfies (9), so does $P(\rho, \theta - \tau)$.   □

A twice continuously differentiable function which satisfies Laplace's equation is called a *harmonic function*. Properties (ii) and (iii) of the Poisson kernel together constitute the "*property of the δ-function.*"[1]

We now state and solve the famous *Dirichlet problem* for the unit disc, which is the following:

Given a continuous function $\varphi(\theta)$ with period 2π, we seek a harmonic function $u(\rho e^{i\theta})$ on the open unit disc which satisfies

$$\lim_{\rho \to 1} u(\rho e^{i\theta}) = \varphi(\theta). \tag{10}$$

We divide the solution of the problem into several steps.

(a) *Mean value formula*: If on the unit disc the function $u(\rho e^{i\theta})$ is harmonic and is continuous up to the unit circle (i.e. u is continuous on the closed unit disc $|z| \leqslant 1$), then

$$\frac{1}{2\pi} \int_0^{2\pi} u(\rho e^{i\theta})\, d\theta = u(0), \qquad 0 \leqslant \rho \leqslant 1. \tag{11}$$

The proof of this is as follows: From the Laplace equation we know that for $\rho > 0$,

$$\rho \frac{\partial}{\partial \rho} \left(\rho \frac{\partial}{\partial \rho} \cdot \frac{1}{2\pi} \int_0^{2\pi} u(\rho e^{i\theta})\, d\theta \right) = -\frac{1}{2\pi} \int_0^{2\pi} \frac{\partial^2}{\partial \theta^2} u(\rho e^{i\theta})\, d\theta$$

$$= -\frac{1}{2\pi} \frac{\partial}{\partial \theta} u(\rho e^{i\theta}) \Big|_0^{2\pi} = 0.$$

Integrating, we have

$$\rho \frac{\partial}{\partial \rho} \cdot \frac{1}{2\pi} \int_0^{2\pi} u(\rho e^{i\theta})\, d\theta = k,$$

---

[1] Translator's note: in the language of Schwarz's theory of distribution, $\lim_{\rho \to 1} P(\rho, \theta)$ is the kernel which represents the Dirac δ-measure on the unit circle.

and when $\rho \to 0$, it can be seen that $k = 0$. Integrating again, we have

$$\frac{1}{2\pi} \int_0^{2\pi} u(\rho e^{i\theta})\, d\theta = c,$$

which is a constant independent of ρ. Once again letting $\rho \to 0$, we obtain equation (11). $\qquad\square$

(b) Using the change of variables of step (a), let

$$v(z) = u(w),$$

so that

$$v(e^{i\tau}) = u(e^{i\psi}), \qquad v(a) = u(0).$$

For $\rho = 1$, equation (11) becomes

$$v(a) = u(0) = \frac{1}{2\pi} \int_0^{2\pi} u(e^{i\psi})\, d\psi$$

$$= \frac{1}{2\pi} \int_0^{2\pi} v(e^{i\tau}) \frac{1 - |a|^2}{|1 - \bar{a}e^{i\tau}|^2}\, d\tau.$$

Then by letting $a = \rho e^{i\theta}$ and changing the symbol, we obtain the *Poisson formula*

$$u(\rho e^{i\theta}) = \frac{1}{2\pi} \int_0^{2\pi} u(e^{i\tau}) \frac{1 - \rho^2}{1 - 2\rho \cos(\theta - \tau) + \rho^2}\, d\tau. \qquad (12)$$

In other words, if $u(\rho e^{i\theta})$ is a harmonic function, then the above formula holds.

(c) *Maximum (minimum) principle.* A nonconstant function, harmonic in the open unit disc and continuous up to the unit circle, must assume its maximum (minimum) on the unit circle.

PROOF. Suppose for some $\rho < 1$, $u(\rho e^{i\theta})$ is the maximum value, then from (12),

$$\frac{1}{2\pi} \int_0^{2\pi} u(e^{i\tau}) \frac{1 - \rho^2}{1 - 2\rho \cos(\theta - \tau) + \rho^2}\, d\tau$$

$$\leqslant u(\rho e^{i\theta}) \frac{1}{2\pi} \int_0^{2\pi} \frac{(1 - \rho^2)\, d\tau}{1 - 2\rho \cos(\theta - \tau) + \rho^2} = u(\rho e^{i\theta}).$$

If u is nonconstant, then there is an arc on the unit circle on which $u < u(\rho e^{i\theta})$. Then the above inequality is a strict inequality, contradicting the Poisson formula. Similarly, the minimum is assumed on the unit circle. $\qquad\square$

(d) *Uniqueness of the solution to the Dirichlet problem:* Suppose there exist two solutions, $u(\rho e^{i\theta})$ and $v(\rho e^{i\theta})$, both satisfying (10), then

$$w(\rho e^{i\theta}) = u(\rho e^{i\theta}) - v(\rho e^{i\theta})$$

is also a harmonic function. Since this harmonic function equals 0 on the unit circle, $w(e^{i\theta}) = 0$. From (c) we know that on the closed disc $|z| \leq 1$, the maximum value of $w(\rho e^{i\theta})$ is ≤ 0 and the minimum value is ≥ 0, whence $w \equiv 0$. It follows that the solution is unique.

(e) *Existence of the solution*: Consider the *Poisson integral*

$$u(\rho e^{i\theta}) = \frac{1}{2\pi} \int_0^{2\pi} P(\rho, \theta - \tau)\varphi(\tau)\, d\tau. \tag{13}$$

The function has the following properties: First of all, it can be seen from (iv) that on the unit disc, $u(\rho e^{i\theta})$ satisfies the Laplace equation, and secondly, from the "δ-function" properties (ii) and (iii), (10) may be proved as follows. From (iii)

$$\varphi(\theta) = \frac{1}{2\pi} \int_0^{2\pi} P(\rho, \theta - \tau)\varphi(\theta)\, d\tau.$$

Now, since $\varphi(\theta)$ is a continuous function, given ε, there exists δ such that for $|\theta - \tau| < \delta$,

$$|\varphi(\theta) - \varphi(\tau)| < \varepsilon. \tag{14}$$

Dividing the integral

$$u(\rho e^{i\theta}) \quad \varphi(\theta) = \frac{1}{2\pi} \int_0^{2\pi} P(\rho, \theta - \tau)(\varphi(\tau) - \varphi(\theta))\, d\tau$$

into two parts, according to (14), we obtain on the one hand:

$$\left| \frac{1}{2\pi} \int_{|\theta - \tau| < \delta} P(\rho, \theta - \tau)(\varphi(\tau) - \varphi(\theta))\, d\tau \right| \leq \varepsilon \cdot \frac{1}{2\pi} \int_0^{2\pi} P(\rho, \theta - \tau)\, d\tau = \varepsilon.$$

On the other hand, when $|\theta - \tau| \geq \delta$, we may use (ii) to choose ρ sufficiently close to 1 such that

$$P(\rho, \theta - \tau) < \varepsilon/2M,$$

where M is the upper bound of $|\varphi(\tau)|$. Consequently

$$\left| \frac{1}{2\pi} \int_{|\theta - \tau| \geq \delta} P(\rho, \theta - \tau)(\varphi(\tau) - \varphi(\theta))\, d\tau \right| < 2M \cdot \frac{1}{2\pi} \int_0^{2\pi} \frac{\varepsilon}{2M}\, d\tau = \varepsilon.$$

Combining the two preceding expressions, we have that for ρ sufficiently close to 1,

$$|u(\rho e^{i\theta}) - \varphi(\theta)| < 2\varepsilon,$$

whence

$$\lim_{\rho \to 1} u(\rho e^{i\theta}) = \varphi(\theta).$$

Therefore expression (13) solves the existence part of the Dirichlet problem on the unit disc.

1.2 Real Forms

In order to see that the foregoing is capable of generalizations, we first look at the "real forms" of the results in §1. Consider first the real form of (1.1). Write

$$w = \frac{z - a}{1 - \bar{a}z} = \frac{(z - a)(1 - a\bar{z})}{(1 - \bar{a}z)(1 - a\bar{z})} = \frac{z - a - az\bar{z} + a^2\bar{z}}{1 - \bar{a}z - a\bar{z} + a\bar{a}z\bar{z}}.$$

Introduce now the notation: for each complex number $v = \xi + i\eta$, denote by v^* the vector (ξ, η), which we also identify with the row matrix $[\xi \; \eta]$. Then clearly for all complex numbers a, b, we have

$$a\bar{b} + \bar{a}b = 2a^*b^{*\prime},$$

where the prime on b^* denotes transpose, and the right side denotes matrix multiplication.

Furthermore, for $a = b + ic$ and $z = x + iy$, we have

$$a^2\bar{z} = (b^2 - c^2 + 2bci)(x - iy),$$

so that

$$(a^2\bar{z})^* = [(b^2 - c^2)x + 2bcy, \; 2bcx - (b^2 - c^2)y]$$

$$= (x, y)\begin{pmatrix} b^2 - c^2 & 2bc \\ 2bc & -b^2 + c^2 \end{pmatrix}$$

$$= (x, y)[2(b, c)'(b, c) - (b, c)(b, c)'I]$$

$$= z^*(2a^{*\prime}a^* - a^*a^{*\prime}I),$$

where I generically stands for the identity matrix (in this case the 2×2 identity matrix). Applying matrix multiplication in like manner to the above expression of w, we obtain

$$w^* = \frac{z^* - a^* - z^*z^{*\prime}a^* + z^*(2a^{*\prime}a^* - a^*a^{*\prime}I)}{1 - 2a^*z^{*\prime} + a^*a^{*\prime}z^*z^{*\prime}}.$$

1.3 The Geometry of the Unit Ball

The real form of the preceding section suggests the following possible generalization:

Let $x = (x_1, \ldots, x_n)$ represent an n-dimensional vector, so that the set of all x such that

$$xx' < 1 \tag{1}$$

represents the unit ball in n-space; here x is identified with the row vector $[x_1 \cdots x_n]$, x' denotes its transpose and xx' denotes matrix multiplication (alternately, xx' may be interpreted as the inner product of x with itself). The last formula of §2 suggests that the transformation

$$y = \frac{x - a - xx'a + x(2a'a - aa'I)}{1 - 2ax' + aa'xx'}, \qquad aa' < 1 \qquad (2)$$

could be a one-to-one transformation of the unit ball onto itself which transforms $x = a$ into $y = 0$. We now prove this.

First rewrite y as

$$y = \frac{(1 - aa')(x - a) - a(x - a)(x - a)'}{1 - 2ax' + aa'xx'}, \qquad aa' < 1. \qquad (3)$$

Then taking the inner product, we get

$$yy' = \frac{(1 - aa')^2(x - a)(x - a)'}{(1 - 2ax' + aa'xx')^2}$$

$$- \frac{2(1 - aa')(x - a)(x - a)'a(x - a)'}{(1 - 2ax' + aa'xx')^2}$$

$$+ \frac{aa'[(x - a)(x - a)']^2}{(1 - 2ax' + aa'xx')^2}$$

$$= \frac{(x - a)(x - a)'}{(1 - 2ax' + aa'xx')^2} [(1 - aa')^2$$

$$- 2(1 - aa')(x - a)a' + aa'(x - a)(x - a)']$$

$$= \frac{(x - a)(x - a)'}{1 - 2ax' + aa'xx'}, \qquad (4)$$

From (3) we know that

$$y + yy'a = \frac{(1 - aa')(x - a)}{1 - 2ax' + aa'xx'}, \qquad (5)$$

Again taking the inner product yields

$$(y + yy'a)(y + yy'a)' = \frac{(1 - aa')^2(x - a)(x - a)'}{(1 - 2ax' + aa'xx')^2}.$$

(4) gives

$$yy'(1 + 2ay' + aa'yy') = \frac{(1 - aa')^2 yy'}{1 + 2ax' + aa'xx'}.$$

If $yy' = 0$ then $y = 0$, and from (4), $x = a$. If $yy' \neq 0$, then we have the equality

$$1 + 2ay' + aa'yy' = \frac{(1 - aa')^2}{1 - 2ax' + aa'xx'} \tag{6}$$

(which still holds for $y = 0$, $x = a$). Substituting into (5), we obtain

$$x = a + \frac{(y + yy'a)(1 - aa')}{1 + 2ay' + aa'yy'},$$

so that

$$x = \frac{y + a + ayy' + y(2a'a - aa'I)}{1 + 2ay' + aa'yy'}. \tag{7}$$

The preceding equation is formally identical with (2), except that a has been replaced by $-a$. Therefore (2) is indeed a one-to-one transformation defined on the unit ball. (This in fact holds for all x, except for the point x which makes the denominator of (2) vanish. It is not difficult to prove that the corresponding exceptional value is $y = -a/aa'$.)

Again from (4) we have

$$1 - yy' = \frac{1 - 2ax' + aa'xx' - (x - a)(x - a)'}{1 - 2ax' + aa'xx'}$$

$$= \frac{(1 - aa')(1 - xx')}{1 - 2ax' + aa'xx'}. \tag{8}$$

As for the denominator, the Schwarz inequality gives us

$$1 - 2ax' - aa'xx' = (1 - ax')^2 + aa'xx' - (ax')^2 > 0.$$

This proves that $1 - yy' > 0$, so that $yy' < 1$. Thus (2) transforms the unit ball into itself. Again from (7), we see that indeed (2) is a one-to-one transformation of the unit ball *onto* itself.

Besides (2) there is also the transformation defined by an orthogonal matrix Γ such that

$$y = x\Gamma, \qquad \Gamma\Gamma' = I. \tag{9}$$

This clearly also transforms the unit ball onto itself.

The group generated by (2) and (9) is the group we wish to discuss. For now we must study the geometric properties of the unit ball which are invariant under this group.

This space is called the *hyperbolic space*, and the group generated by (2) and (9) is called the *group of non-Euclidean motions*.

Under the action of this group, any point in the unit ball can be transformed into the origin; furthermore, any n mutually orthogonal directions can be transformed into the n axes of a rectangular coordinate system.

1.4 The Differential Metric

We seek the differential of the transformation

$$y = \frac{(1 - aa')(x - a) - (x - a)(x - a)'a}{1 - 2ax' + aa'xx'}$$

A straightforward computation gives:

$$
\begin{aligned}
(1 - 2ax' + aa'xx')^2\, dy &= (1 - 2ax' + aa'xx')[(1 - aa')\,dx - 2\,dx\,(x - a)'a] \\
&\quad - [-2\,dxa' + 2aa'\,dxx'][(1 - aa')(x - a) \\
&\quad - (x - a)(x - a)'a] \\
&= (1 - aa')\,dx\,\{(1 - 2ax' + aa'xx')I - 2(1 - 2ax')x'a \\
&\quad + 2a'x - 2xx'a'a - 2aa'x'x\} \\
&= (1 - aa')\,dx\,\{(1 - 2ax' + aa'xx')I - 2(1 - ax') \\
&\quad \times (x'a - a'x) + 2(x'a - a'x)^2\}.
\end{aligned}
$$

Let

$$P = (1 - 2ax' + aa'xx')I - 2(1 - ax')(x'a - a'x) + 2(x'a - a'x)^2.$$

By further defining

$$M = x'a - a'x, \qquad \lambda = 1 - 2ax' + aa'xx', \tag{1}$$

we can write

$$P = \lambda I - 2(1 - ax')M + 2M^2 \tag{2}$$

as well as

$$dy = \frac{1 - aa'}{(1 - 2ax' + aa'xx')^2}\, dxP. \tag{3}$$

It can be easily proved that

$$
\begin{aligned}
xM^2 &= [(ax')^2 - aa'xx']x, \\
aM^2 &= [(ax')^2 - aa'xx']a, \\
M^3 &= [(ax')^2 - aa'xx']M,
\end{aligned}
\tag{4}
$$

whence

$$
\begin{aligned}
PP' &= (\lambda I - 2(1 - ax')M + 2M^2)(\lambda I + 2(1 - ax')M + 2M^2) \\
&= (\lambda I + 2M^2)^2 - 4(1 - ax')^2 M^2 \\
&= \lambda^2 I + 4(\lambda - (1 - ax')^2)M^2 + 4M^4 \\
&= \lambda^2 I + 4M\{[aa'xx' - (ax')^2]M + M^3\} \\
&= \lambda^2 I.
\end{aligned}
\tag{5}
$$

Thus

$$dy\,dy' = \frac{(1 - aa')^2}{(1 - 2ax' + aa'xx')^4}\,dxPP'\,dx'$$

$$= \frac{(1 - aa')^2}{(1 - 2ax' + aa'xx')^2}\,dx\,dx', \tag{6}$$

which, together with (3.8), immediately gives

$$\frac{dy\,dy'}{(1 - yy')^2} = \frac{dx\,dx'}{(1 - xx')^2}. \tag{7}$$

This relation is also invariant under transformation (3.9). Therefore (7) is an invariant differential expression of degree 2.

1.5 A Differential Operator

We will now prove that the partial differential equation

$$(1 - yy')^2 \sum_{i=1}^{n} \frac{\partial^2 u}{\partial y_i^2} + 2(n - 2)(1 - yy') \sum_{i=1}^{n} y_i \frac{\partial u}{\partial y_i} = 0 \tag{1}$$

is invariant under transformation (3.2). Since (1) may be rewritten as

$$(1 - yy')^n \sum_{i=1}^{n} \frac{\partial}{\partial y_i}\left[(1 - yy')^{2-n} \frac{\partial u}{\partial y_i}\right] = 0,$$

it is equivalent to proving

$$(1 - yy')^n \sum_{i=1}^{n} \frac{\partial}{\partial y_i}\left[(1 - yy')^{2-n} \frac{\partial u}{\partial y_i}\right] = (1 - xx')^n \sum_{i=1}^{n} \frac{\partial}{\partial x_i}\left[(1 - xx')^{2-n} \frac{\partial u}{\partial x_i}\right], \tag{2}$$

where x and y are related as in (3.2).

Before proving that (2) holds, let us first prove the following lemmas.

Lemma 1. *Let $\mu = 1 + 2ay' + aa'yy'$, then*

$$\sum_{i=1}^{n} \frac{\partial}{\partial y_i}\left(\frac{1}{\mu^{n-2}} \frac{\partial x_k}{\partial y_i}\right) = 0. \tag{3}$$

PROOF. From (3.7) we already know that

$$x_k = a_k + \frac{(1 - aa')(y_k + yy'a_k)}{1 + 2ay' + aa'yy'}.$$

Therefore it suffices to prove

$$\sum_{i=1}^{n} \frac{\partial}{\partial y_i} \frac{1}{\mu^{n-2}} \frac{\partial}{\partial y_i} \frac{y_k + yy'a_k}{\mu} = 0. \tag{4}$$

The left side equals

$$\sum_{i=1}^{n} \frac{\partial}{\partial y_i} \frac{1}{\mu^{n}} [(\delta_{ik} + 2y_i a_k)(1 + 2ay' + aa'yy') - 2(y_k + yy'a_k)(a_i + aa'y_i)]$$

$$= \frac{1}{\mu^{n+1}} \sum_{i=1}^{n} \{[2a_k(1 + 2ay' + aa'yy') + 2(\delta_{ik} + 2y_i a_k)(a_i + aa'y_i)$$

$$- 2(\delta_{ik} + 2y_i a_k)(a_i + aa'y_i) - 2(y_k + yy'a_k)aa'](1 + 2ay' + aa'yy')$$

$$- 2n[(\delta_{ik} + 2y_i a_k)(1 + 2ay' + aa'yy')$$

$$- 2(y_k + yy'a_k)(a_i + aa'y_i)](a_i + aa'y_i)\}$$

$$= \frac{1}{\mu^{n+1}} \sum_{i=1}^{n} \{[2a_k(1 + 2ay') - 2y_k aa'](1 + 2ay' + aa'yy')$$

$$- 2n(\delta_{ik} + 2y_i a_k)(a_i + aa'y_i)(1 + 2ay' + aa'yy')$$

$$+ 4n(y_k + yy'a_k)(a_i + aa'y_i)(a_i + aa'y_i)\}$$

$$= \frac{1}{\mu^{n}} \{n[2a_k(1 + 2ay') - 2y_k aa'] - 2n(a_k + aa'y_k + 2ay'a_k + 2aa'yy'a_k)$$

$$+ 4n(y_k + yy'a_k)aa'\}$$

$$= 0. \qquad \square$$

Lemma 2.

$$\sum_{i=1}^{n} \frac{\partial x_j}{\partial y_i} \frac{\partial x_k}{\partial y_i} = \frac{\lambda^2}{(1 - aa')^2} \delta_{jk}.$$

PROOF. This is very easily derived from

$$dy\, dy' = \frac{(1 - aa')^2}{\lambda^2} dx\, dx'. \qquad \square$$

Now let us go back and prove (2). From (3.8), it is immediate that

$$\left(\frac{(1 - aa')(1 - xx')}{\lambda(x)}\right)^n \sum_{i=1}^{n} \sum_{j=1}^{n} \frac{\partial}{\partial x_j} \left[\left(\frac{(1 - aa')(1 - xx')}{\lambda(x)}\right)^{2-n} \sum_{k=1}^{n} \frac{\partial u}{\partial x_k} \frac{\partial x_k}{\partial y_i}\right] \frac{\partial x_j}{\partial y_i}$$

$$= (1 - aa')^2 \left(\frac{1 - xx'}{\lambda(x)}\right)^n \sum_{i,j,k} \frac{\partial}{\partial x_j} \left[(1 - xx')^{2-n} \frac{\partial u}{\partial x_k}\right] \lambda^{n-2} \frac{\partial x_k}{\partial y_i} \frac{\partial x_j}{\partial y_i}$$

$$+ (1 - aa')^2 \left(\frac{1 - xx'}{\lambda(x)}\right)^n \sum_{i,j,k} (1 - xx')^{2-n} \frac{\partial u}{\partial x_k} \frac{\partial}{\partial x_j} \left(\lambda^{n-2} \frac{\partial x_k}{\partial y_i}\right) \frac{\partial x_j}{\partial y_i}$$

$$\equiv s_1 + s_2.$$

From lemma 2, s_1 is just the right side of (2). Thus it remains to be shown that $s_2 = 0$, that is

$$\sum_{i,j,k} \frac{\partial u}{\partial x_k} \frac{\partial}{\partial x_j} \left(\lambda^{n-2} \frac{\partial x_k}{\partial y_i} \right) \frac{\partial x_j}{\partial y_i} = 0.$$

But this equality is easily deduced from the identity

$$\sum_{i,j} \frac{\partial}{\partial x_j} \left(\lambda^{n-2} \frac{\partial x_k}{\partial y_i} \right) \frac{\partial x_j}{\partial y_i} = 0,$$

which is implied by

$$\sum_{i=1}^{n} \frac{\partial}{\partial y_i} \left(\lambda^{n-2} \frac{\partial x_k}{\partial y_i} \right) = 0.$$

In turn, the latter may be proved as follows: from (3.6) we know that $\lambda \mu = (1 - aa')^2$, therefore the above equality follows from lemma 1. □

1.6 Spherical Coordinates

Let

$$x = \rho u, \qquad uu' = 1,$$

where $0 \leqslant \rho < \infty$ and u is a vector in n-space, then $du \cdot u' = 0$. Thus

$$dx\, dx' = (d\rho u + \rho\, du)(d\rho u + \rho\, du)'$$
$$= d\rho^2 + \rho^2\, du\, du'.$$

And so we have

$$\frac{dx\, dx'}{(1 - xx')^2} = \frac{d\rho^2 + \rho^2\, du\, du'}{(1 - \rho^2)^2}. \tag{1}$$

Now we introduce spherical coordinates

$$u = (\cos \theta_1, \sin \theta_1 \cos \theta_2, \sin \theta_1 \sin \theta_2 \cos \theta_3, \ldots,$$
$$\sin \theta_1 \cdots \sin \theta_{n-2} \cos \theta_{n-1}, \sin \theta_1 \cdots \sin \theta_{n-2} \sin \theta_{n-1}),$$
$$0 \leqslant \theta_1, \theta_2, \ldots, \theta_{n-2} \leqslant \pi, 0 \leqslant \theta_{n-1} \leqslant 2\pi.$$

The unit circle may be expressed as $(\cos \theta, \sin \theta)$, $0 \leqslant \theta \leqslant 2\pi$, but this is not to say that the unit circle is equivalent to the interval $[0, 2\pi]$, for the continuous functions on the interval $[0, 2\pi]$ are not necessarily continuous functions on the unit circle. The reason for this is that $\theta = 0$ and $\theta = 2\pi$ in fact represent the same point on the unit circle. Thus when we speak of a continuous function $f(\theta)$ on the unit circle, we must keep in mind that it is a function of period 2π.

On the unit sphere, the situation is even more complicated: since $(\cos \theta_1, \sin \theta_1 \cos \theta_2, \ldots, \sin \theta_1 \sin \theta_2 \cdots \sin \theta_{n-2} \sin \theta_{n-1})$ is not a function of θ_1 of period 2π, then how would we define continuity on the sphere for the function

$$f(\theta_1, \ldots, \theta_{n-1}), \qquad 0 \leqslant \theta_1, \ldots, \theta_{n-2} \leqslant \pi, \qquad 0 \leqslant \theta_{n-1} \leqslant 2\pi?$$

The main point here is to observe the behavior at the endpoints of the interval. Let us first consider the points represented by $\theta_1 = 0$, putting aside for the time being the variables $\theta_2, \ldots, \theta_{n-1}$. When $\theta_1 = 0$, then $u = e_1 \equiv (1, 0, \ldots, 0)$. Thus for $u = e_1$, the continuous function $f(\theta_1, \ldots, \theta_{n-1})$ on the sphere necessarily gives us the existence of

$$\lim_{\theta_1 \to +0} f(\theta_1, \ldots, \theta_{n-1}),$$

and furthermore this is independent of $\theta_2, \ldots, \theta_{n-1}$. By the same reasoning,

$$\lim_{\theta_1 \to \pi - 0} f(\theta_1, \ldots, \theta_{n-1})$$

is the value of the function at $u = -e_1$, and again this is independent of $\theta_2, \ldots, \theta_{n-1}$. Continuing in the same manner we see that

$$\lim_{\theta_2 \to +0} f(\theta_1, \theta_2, \ldots, \theta_{n-1})$$

is independent of $\theta_1, \theta_3, \ldots, \theta_{n-1}$. Finally, we also have

$$\lim_{\theta_{n-1} \to +0} f(\theta_1, \ldots, \theta_{n-1}) = \lim_{\theta_{n-1} \to 2\pi - 0} f(\theta_1, \ldots, \theta_{n-1}).$$

The only continuous functions on the sphere are exactly those satisfying these conditions.

Differentiating the vectorial quantity u, we can easily deduce:

$$du \, du' = d\theta_1^2 + \sin^2 \theta_1 \, d\theta_2^2 + \sin^2 \theta_1 \sin^2 \theta_2 \, d\theta_3^2 + \cdots$$
$$+ \sin^2 \theta_1 \cdots \sin^2 \theta_{n-2} \, d\theta_{n-1}^2. \qquad (2)$$

On the sphere, the volume element $\dot{u}$ is the square root of the determinant of this quadratic differential form, i.e.

$$\dot{u} = \sin^{n-2} \theta_1 \sin^{n-3} \theta_2 \cdots \sin \theta_{n-2} \, d\theta_1 \, d\theta_2 \cdots d\theta_{n-1}.$$

It is now not difficult to see that the volume of the (unit) sphere is

$$\omega_{n-1} = \frac{2\pi^{n/2}}{n \cdot \Gamma\left(\dfrac{n}{2}\right)}.$$

By a fairly simple and direct calculation, we know that the Laplacian

$$\Delta = \frac{\partial^2}{\partial x_1^2} + \cdots + \frac{\partial^2}{\partial x_n^2} \qquad (3)$$

in polar coordinates is

$$\Delta = \frac{1}{\rho^{n-1}} \frac{\partial}{\partial \rho}\left(\rho^{n-1} \frac{\partial}{\partial \rho}\right) + \frac{1}{\rho^2} \partial_u^2, \qquad (4)$$

where

$$\partial_u^2 = \frac{\partial^2}{\partial\theta_1^2} + \frac{1}{\sin^2\theta_1}\frac{\partial^2}{\partial\theta_2^2} + \cdots + \frac{1}{\sin^2\theta_1\cdots\sin^2\theta_{n-2}}\frac{\partial^2}{\partial\theta_{n-1}^2}$$

$$+ (n-2)\operatorname{ctg}\theta_1\frac{\partial}{\partial\theta_1} + (n-3)\frac{\operatorname{ctg}\theta_2}{\sin^2\theta_1}\frac{\partial}{\partial\theta_2}$$

$$+ (n-4)\frac{\operatorname{ctg}\theta_3}{\sin^2\theta_1\sin^2\theta_2}\frac{\partial}{\partial\theta_3} + \cdots$$

$$+ \frac{\operatorname{ctg}\theta_{n-2}}{\sin^2\theta_1\cdots\sin^2\theta_{n-3}}\frac{\partial}{\partial\theta_{n-2}}. \tag{5}$$

Now we consider the polar coordinate form of the differential operator

$$(1-xx')^2\sum_{i=1}^{n}\frac{\partial^2}{\partial x_i^2} + 2(n-2)(1-xx')\sum_{i=1}^{n}x_i\frac{\partial}{\partial x_i}.$$

From

$$\sum_{i=1}^{n}x_i\frac{\partial}{\partial x_i} = \rho\frac{\partial}{\partial\rho}$$

and (4) we have

$$(1-xx')^2\sum_{i=1}^{n}\frac{\partial^2}{\partial x_i^2} + 2(n-2)(1-xx')\sum_{i=1}^{n}x_i\frac{\partial}{\partial x_i}$$

$$= (1-\rho^2)^2\left[\frac{1}{\rho^{n-1}}\frac{\partial}{\partial\rho}\left(\rho^{n-1}\frac{\partial}{\partial\rho}\right) + \frac{1}{\rho^2}\partial_u^2\right] + 2(n-2)(1-\rho^2)\rho\frac{\partial}{\partial\rho}$$

$$= (1-\rho^2)^2\frac{\partial^2}{\partial\rho^2} + \frac{1-\rho^2}{\rho}[(n-1)+(n-3)\rho^2]\frac{\partial}{\partial\rho} + \frac{(1-\rho^2)^2}{\rho^2}\partial_u^2$$

$$= \frac{(1-\rho^2)^n}{\rho^{n-1}}\frac{\partial}{\partial\rho}\left(\frac{\rho^{n-1}}{(1-\rho^2)^{n-2}}\frac{\partial}{\partial\rho}\right) + \frac{(1-\rho^2)^2}{\rho^2}\partial_u^2. \tag{6}$$

Let us consider the case where (6) is applied to a function $\Phi(\rho\cos\theta_1, \rho\sin\theta_2)$ which is independent of $\theta_2, \theta_3, \ldots, \theta_{n-1}$. We obtain the partial differential equation

$$\frac{(1-\rho^2)^n}{\rho^{n-1}}\frac{\partial}{\partial\rho}\left(\frac{\rho^{n-1}}{(1-\rho^2)^{n-2}}\frac{\partial}{\partial\rho}\right)\Phi + \frac{(1-\rho^2)^2}{\rho^2}\left(\frac{\partial^2}{\partial\theta_1^2} + (n-2)\operatorname{ctg}\theta_1\frac{\partial}{\partial\theta_1}\right)\Phi = 0.$$

Letting $\xi = \cos\theta_1$ yields

$$\frac{(1-\rho^2)^n}{\rho^{n-1}}\frac{\partial}{\partial\rho}\left(\frac{\rho^{n-1}}{(1-\rho^2)^{n-2}}\frac{\partial}{\partial\rho}\right)\Phi + \frac{(1-\rho^2)^2}{\rho^2}\left[(1-\xi^2)\frac{\partial^2}{\partial\xi^2} - (n-1)\xi\frac{\partial}{\partial\xi}\right]\Phi = 0, \tag{7}$$

which may also be written as

$$\left[\frac{(1 - \rho^2)^{n-2}}{\rho^{n-3}} \frac{\partial}{\partial \rho} \left(\frac{\rho^{n-1}}{(1 - \rho^2)^{n-2}} \frac{\partial}{\partial \rho} \right) \right.$$

$$\left. + (1 - \xi^2)^{-(n-3)/2} \frac{\partial}{\partial \xi} \left((1 - \xi^2)^{(n-1)/2} \frac{\partial}{\partial \xi} \right) \right] \Phi = 0. \quad (8)$$

This suggests that in the interior of the rectangle

$$0 \leqslant \rho \leqslant 1, \qquad -1 \leqslant \xi \leqslant 1,$$

we study the following *quasi-conformal* transformation

$$u = u(\rho, \xi),$$
$$v = v(\rho, \xi).$$

This pair of functions u, v, satisfies the system of differential equations

$$\begin{cases} \dfrac{\rho^{n-1}}{(1 - \rho^2)^{n-2}} \dfrac{\partial u}{\partial \rho} = (1 - \xi^2)^{-(n-3)/2} \dfrac{\partial v}{\partial \xi}, \\[2mm] \dfrac{(1 - \rho^2)^{n-2}}{\rho^{n-3}} \dfrac{\partial v}{\partial \rho} = -(1 - \xi^2)^{(n-1)/2} \dfrac{\partial u}{\partial \xi}. \end{cases} \quad (9)$$

From (9) and making use of $\partial^2 v / \partial \xi \, \partial \rho = \partial^2 v / \partial \rho \, \partial \xi$ to cancel v, we see that u satisfies differential equation (8). Using $\partial^2 u / \partial \xi \, \partial \rho = \partial^2 u / \partial \rho \, \partial \xi$ to cancel u, the differential equation satisfied by v is:

$$\left[\frac{\rho^{n-1}}{(1 - \rho^2)^{n-2}} \frac{\partial}{\partial \rho} \left(\frac{(1 - \rho^2)^{n-2}}{\rho^{n-3}} \frac{\partial}{\partial \rho} \right) \right.$$

$$\left. + (1 - \xi^2)^{(n-1)/2} \frac{\partial}{\partial \xi} \left((1 - \xi^2)^{-(n-3)/2} \frac{\partial}{\partial \xi} \right) \right] v = 0.$$

In these two differential equations the second order terms are equal and the sum of the linear terms is equal to

$$\frac{\partial v}{\partial \rho} \cdot \frac{\partial}{\partial \rho} \left(\frac{\rho^{n-1}}{(1 - \rho^2)^{n-2}} \cdot \frac{(1 - \rho^2)^{n-2}}{\rho^{n-3}} \right) + \frac{\partial v}{\partial \xi} \cdot$$

$$\frac{\partial}{\partial \xi} (1 - \xi^2)^{[(n-1)/2] - [(n-3)/2]} = 2\rho \frac{\partial v}{\partial \rho} - 2\xi \frac{\partial v}{\partial \xi}.$$

Expression (9) is perhaps the quasi-conformal transformation closest to an ordinary conformal transformation. Thus a deeper study of this special case might serve as a good reference for the investigation of general quasi-conformal transformations.

1.7 The Poisson Formula

From (4.6), i.e.

$$dy\,dy' = \left(\frac{1 - aa'}{1 - 2ax' + aa'xx'}\right)^2 dx\,dx',$$

it may be seen that on the corresponding spheres defined by $x = u$, $y = v$ and $uu' = vv' = 1$, we also have

$$dv\,dv' = \left(\frac{1 - aa'}{1 - 2au' + aa'uu'}\right)^2 du\,du'.$$

It follows that the volume elements $\dot{u}$ and $\dot{v}$ of the spheres obey the following relation

$$\dot{v} = \left(\frac{1 - aa'}{1 - 2au' + aa'}\right)^{n-1} \dot{u}, \tag{1}$$

This suggests the following possible *Poisson formula*:

$$\Phi(x) = \frac{1}{\omega_{n-1}} \int \cdots \int_{uu'=1} \left(\frac{1 - xx'}{1 - 2xu' + xx'}\right)^{n-1} \Phi(u)\dot{u} \tag{2}$$

What is suggested here is that if on a unit sphere we are given the function $\Phi(u)$, then from (2) we may define a function in the unit ball which in addition to satisfying (2), also satisfies partial differential equation (5.1) inside the ball. Before discussing this problem in detail, let us first study the properties of the Poisson kernel

$$P(x, u) = \left(\frac{1 - xx'}{1 - 2xu' + xx'}\right)^{n-1}. \tag{3}$$

Let $x = \rho v$, then

$$P(x, u) = \left(\frac{1 - \rho^2}{1 - 2\rho \cos \langle u, v \rangle + \rho^2}\right)^{n-1}.$$

Here, $\langle u, v \rangle$ denotes the angle subtended by the two unit vectors u, v.

(a) When $0 \leqslant \rho < 1$, $P(x, u) > 0$. This is obvious since

$$1 - 2\rho \cos \langle u, v \rangle + \rho^2 \geqslant 1 - 2\rho + \rho^2 = (1 - \rho)^2.$$

(b) We have

$$\lim_{\rho \to 1} P(x, u) = \begin{cases} 0, & u \neq v \\ \infty, & u = v. \end{cases}$$

Somewhat more concretely, suppose we let $\langle u, v \rangle = \alpha$ and let δ be a positive number. Then for $|\alpha| > \delta$, given any ε, there exists a ρ_0 such that for $1 > \rho > \rho_0$,

$$P(x, u) \leqslant \left(\frac{1 - \rho^2}{1 - 2\rho \cos \delta + \rho^2}\right)^{n-1} < \varepsilon.$$

(c)
$$\frac{1}{\omega_{n-1}} \int \cdots \int_u P(x,u)\dot{u} = 1,$$

PROOF. This follows immediately from the obvious fact

$$\frac{1}{\omega_{n-1}} \int \cdots \int_v \dot{v} = 1$$

and from (1). ☐

(d) For $\rho < 1$, $P(x,u)$ is a solution to (5.1).

PROOF. Differentiating $P(x,u)$ yields

$$\frac{\partial P(x,u)}{\partial x_i} = 2(n-1)\frac{(1-xx')^{n-2}}{(1-2ux'+xx')^n}[-2(1-ux')x_i + (1-xx')u_i].$$

As a result,

$$\sum_{i=1}^{n} \frac{\partial}{\partial x_i}\left[(1-xx')^{2-n}\frac{\partial P(x,u)}{\partial x_i}\right]$$

$$= 2(n-1)\sum_{i=1}^{n}\frac{\partial}{\partial x_i}\left[\frac{-2(1-ux')x_i + (1-xx')u_i}{(1-2ux'+xx')^n}\right]$$

$$= 2(n-1)\sum_{i=1}^{n}\left\{\frac{-2(1-ux')+2u_ix_i-2x_iu_i}{(1-2ux'+xx')^n}\right.$$

$$\left. - \frac{n[-2(1-ux')x_i+(1-xx')u_i][-2u_i+2x_i]}{(1-2ux'+xx')^{n+1}}\right\}$$

$$= 2(n-1)\left\{\frac{-2n(1-ux')}{(1-2ux'+xx')^n}\frac{2n[2(1-ux')ux'-2(1-ux')xx'}{}\right.$$

$$\left. \frac{-(1-xx')uu'+(1-xx')ux']}{(1-2ux'+xx')^{n+1}}\right\}.$$

Now since $uu' = 1$, the above expression is equal to 0. Thus $P(x,u)$ indeed satisfies equation (5.1). ☐

Theorem 1. *Suppose $\Phi(u)$ is a continuous function defined on the sphere $uu' = 1$, then the Poisson formula*

$$\Phi(x) = \frac{1}{\omega_{n-1}}\int \cdots \int_{uu'=1}\left(\frac{1-xx'}{1-2xu'+xx'}\right)^{n-1}\Phi(u)\dot{u}$$

defines a function in the unit ball which satisfies equation (5.1). Moreover for all v such that $vv' = 1$,

$$\lim_{r\to 1}\Phi(rv) = \Phi(v).$$

PROOF. When $r < 1$, $P(x, u)$ satisfies (5.1). Thus by differentiating under the integral sign, we know that $\Phi(x)$ also satisfies (5.1).

Now we must prove that when $r \to 1$, $\Phi(rv) - \Phi(v)$ tends to 0. We have

$$\Phi(rv) - \Phi(v) = \frac{1}{\omega_{n-1}} \int \cdots \int_{uu' = 1} \left(\frac{1 - r^2}{1 - 2ruv' + r^2} \right)^{n-1} (\Phi(u) - \Phi(v))\dot{u}.$$

Let $\cos \alpha = vu'$, and divide the integral into:

$$\Phi(rv) = \frac{1}{\omega_{n-1}} \left(\int \cdots \int_{|\alpha| < \delta} + \int \cdots \int_{|\alpha| < \delta} \right) \equiv s_1 + s_2.$$

Then by choosing a δ sufficiently small such that

$$|\Phi(u) - \Phi(v)| < \varepsilon,$$

we get

$$s_1 = O\left(\varepsilon \int \cdots \int_{uu' = 1} \left(\frac{1 - r^2}{1 - 2ruv' + r^2} \right)^{n-1} \dot{u} \right) = O(\varepsilon).$$

Now for the above chosen δ, take r sufficiently close to 1 so that

$$\left| \frac{1 - r^2}{1 - 2ruv' + r^2} \right|^{n-1} \leqslant \varepsilon.$$

Then also

$$s_2 = O(\varepsilon). \qquad \square$$

1.8 What Has the Above Suggested?

What we have discussed above suggests at least three points:

(a) Are there other transitive groups which take the unit ball into itself?
(b) Use the "δ-function" as a starting point, i.e. to put the properties of $P(x, u)$ in an abstract setting for the study of the Dirichlet problem of partial differential equations.
(c) Start with the harmonic analysis of the boundary.

We begin by providing the answers to problems (a) and (b). (c) will be put aside until the next chapter.

(a) Consider the transformation

$$y = \frac{\sqrt{1 - aa'}(x - a)(1 + \lambda a'a)}{1 - ax'}, \tag{1}$$

where $aa' < 1$, and

$$\lambda = \frac{1 - \sqrt{1 - aa'}}{aa' \sqrt{1 - aa'}}, \qquad 1 + \lambda aa' = \frac{1}{\sqrt{1 - aa'}}.$$

Let us first prove that (1) takes the unit ball into itself. Since

$$yy' = \frac{1 - aa'}{(1 - ax')^2} (x - a)(I + \lambda a'a)^2 (x - a)'$$

$$= \frac{1 - aa'}{(1 - ax')^2} (x - a)\left(I + \frac{1}{1 - aa'} a'a\right)(x - a)'$$

$$= \frac{(1 - aa')(x - a)(x - a)' + [(x - a)a']^2}{(1 - ax')^2},$$

therefore

$$1 - yy' = \frac{(1 - aa')[(1 - 2ax' + aa') - (x - a)(x - a)']}{(1 - ax')^2}$$

$$= \frac{(1 - aa')(1 - xx')}{(1 - ax')^2}. \tag{2}$$

It is not difficult to prove that

$$\frac{dy(I - y'y)^{-1} dy'}{1 - yy'} = \frac{dx(I - x'x)^{-1} dx'}{1 - xx'}, \tag{3}$$

i.e.

$$dy(I - y'y)^{-1} dy' = \frac{1 - aa'}{(1 - ax')^2} dx(I - x'x)^{-1} dx'.$$

On the unit sphere $x = u$, $y = v$, since $duu' = 0$, we have

$$dv\, dv' = \frac{1 - aa'}{(1 - au')^2} du\, du'.$$

Hence we obtain the Poisson kernel

$$P(x, u) = \frac{(1 - xx')^{(n-1)/2}}{(1 - ux')^{n-1}}. \tag{4}$$

The differential equation satisfied by $P(x, u)$ is

$$\sum_{i=1}^{n} \frac{\partial^2 \Phi}{\partial x_i^2} - \sum_{i,j=1}^{n} x_i x_j \frac{\partial^2 \Phi}{\partial x_i \partial x_j} - 2 \sum_{i=1}^{n} x_i \frac{\partial \Phi}{\partial x_i} = 0. \tag{5}$$

Thus we have proved that the Poisson formula

$$\Phi(x) = \frac{1}{\omega_{n-1}} \int \cdots \int_{uu'=1} \frac{(1 - xx')^{(n-1)/2}}{(1 - ux')^{n-1}} \Phi(u)\dot{u}, \tag{6}$$

gives a solution to the Dirichlet problem of partial differential equation (5). (Once uniqueness is proved, the Dirichlet problem is completely solved.)

  (b) Beginning not with a group, but only with the "Poisson kernel" is also a possibility.

Let $\mathscr{D}$ be a domain with $\mathscr{L}$ as its boundary, if we could find a function

$$P(x, u),$$

where $x \in \mathscr{D}$ and $u \in \mathscr{L}$, which satisfies the following properties:

(i) $P(x, u) > 0$,
(ii) $\int_{\mathscr{L}} P(x, u)\dot{u} = 1$,
(iii) when x tends toward a boundary point v,

$$\lim_{x \to v} P(x, u) = \begin{cases} 0, & \text{for } u \neq v \\ \infty, & \text{for } u = v, \end{cases}$$

(iv) it satisfies a linear operator equation (not necessarily a differential equation)

$$\partial \Phi = 0, \tag{A}$$

then from

$$\Phi(x) = \int_L P(x, u)\Phi(u)\dot{u}$$

we could hope to find a solution to the Dirichlet problem for equation (A). Take, for example, the function in the unit ball given by

$$\frac{1 - xx'}{(1 - 2ux' + xx')^{n/2}}.$$

It also possesses the following properties:

(1)   nonnegativity
(2), (3) "property of the δ-function"
(4)   it satisfies the ordinary Laplace equation

$$\sum_{i=1}^{n} \frac{\partial^2 \Phi}{\partial x_i^2} = 0.$$

Hence the Poisson integral,

$$\Phi(x) = \frac{1}{\omega_{n-1}} \int \cdots \int_{uu'=1} \frac{1 - xx'}{(1 - 2ux' + xx')^{n/2}} \Phi(u)\dot{u},$$

naturally also yields a solution to the Dirichlet problem on the unit ball. Among the properties (i)–(iv) listed above, there are only two points which are relatively difficult to prove for this function:

(i) $\dfrac{1}{\omega_{n-1}} \displaystyle\int \cdots \int_{uu'=1} \dfrac{1 - xx'}{(1 - 2ux' + xx')^{n/2}} \dot{u} = 1$,

(ii) $\displaystyle\sum_{i=1}^{n} \dfrac{\partial^2}{\partial x_i^2} \left[\dfrac{1 - xx'}{(1 - 2ux' + xx')^{n/2}} \right] = 0.$

Remark. One can find other much more complicated types of "δ-functions" in the book, *Harmonic Analysis of Functions of Several Complex Variables in the Classical Domains*, by L. K. Hua, (tr. by L. Ebner and A. Korányi) Amer. Math. Soc., Providence, 1963.

1.9 The Symmetry Principle

One of the main reasons for the importance of the symmetry principle is the following: the 2-dimensional Laplace equation,

$$\left[\rho \frac{\partial}{\partial \rho}\left(\rho \frac{\partial}{\partial \rho}\right) + \frac{\partial^2}{\partial \theta^2}\right]\Phi = 0$$

is invariant under inversion, i.e. letting

$$\tau = 1/\rho,$$

then

$$\rho \frac{\partial}{\partial \rho} = -\tau \frac{\partial}{\partial \tau}.$$

However, when $n \geqslant 3$, this property no longer holds. The Laplace equation

$$\frac{1}{\rho^{n-3}} \frac{\partial}{\partial \rho}\left(\rho^{n-1} \frac{\partial}{\partial \rho}\right) + \partial_u^2 = \rho^2 \frac{\partial^2}{\partial \rho^2} + (n-1)\rho \frac{\partial}{\partial \rho} + \partial_u^2$$

becomes

$$\frac{1}{\rho^{n-2}}\left(\rho \frac{\partial}{\partial \rho}\right)\left(\rho^{n-2}\rho \frac{\partial}{\partial \rho}\right) + \partial_u^2 = \tau^{n-2}\left(\tau \frac{\partial}{\partial \tau}\right)\left(\frac{1}{\tau^{n-2}}\tau \frac{\partial}{\partial \tau}\right) + \partial_u^2$$

$$= \tau^{n-1} \frac{\partial}{\partial \tau}\left(\tau^{-n+3} \frac{\partial}{\partial \tau}\right) + \partial_u^2$$

$$= \tau^2 \frac{\partial^2}{\partial \tau^2} - (n-3)\tau \frac{\partial}{\partial \tau} + \partial_u^2,$$

that is to say, the Laplace equation is not invariant under inversion. Fortunately we have a replacement: if we rewrite Φ as $\tau^{n-2}\psi$, then

$$\frac{\partial \Phi}{\partial \tau} = \tau^{n-2} \frac{\partial \Psi}{\partial \tau} + (n-2)\tau^{n-3}\Psi,$$

$$\frac{\partial^2 \Phi}{\partial \tau^2} = \tau^{n-2} \frac{\partial^2 \Psi}{\partial \tau^2} + 2(n-2)\tau^{n-3} \frac{\partial \Psi}{\partial \tau} + (n-2)(n-3)\tau^{n-4}\Psi.$$

Consequently

$$\left[\tau^2 \frac{\partial^2}{\partial \tau^2} - (n-3)\tau \frac{\partial}{\partial \tau} + \partial_u^2 \right]\Phi = \tau^{n-2}\left[\tau^2 \frac{\partial^2 \Psi}{\partial \tau^2} + (n-1)\tau \frac{\partial \Psi}{\partial \tau}\right] + \tau^{n-2}\partial_u^2 \Psi,$$

that is, if we subject the variable x and the function Φ to the transformations

$$y = x/xx', \qquad \Psi(y) = (xx')^{n/2-1}\Phi(x),$$

then the Laplace equation is also invariant.

In other words, when studying the symmetry principle of the n-dimensional Laplace equation, we must note that both the variable x and the function Φ must undergo transformation. However, as for the differential equation we have been discussing, viz.

$$\frac{(1-\rho^2)^n}{\rho^{n-1}} \frac{\partial}{\partial \rho}\left(\frac{\rho^{n-1}}{(1-\rho^2)^{n-2}} \frac{\partial \Phi}{\partial \rho}\right) + \frac{(1-\rho^2)^2}{\rho^2} \partial_u^2 \Phi = 0,$$

it is invariant under $\rho = 1/\tau$. Thus we may directly generalize in this case as in the case of two variables.

1.10 The Invariance of the Laplace Equation

The Laplace equation does not remain invariant under

$$y = \frac{(1-aa')(x-a) - (x-a)(x-a)'a}{1 - 2ax' + aa'xx'} \qquad (aa' < 1). \tag{1}$$

However if the differentiated function correspondingly undergoes a transformation, then we may find other invariance properties, that is to say, if an independent variable undergoes transformation according to (1) and the function is transformed according to

$$Y = \left(\frac{1 - 2ax' + aa'xx'}{1 - aa'}\right)^{n/2-1} X, \tag{2}$$

we then have

$$(1 - xx')^{n/2+1} \sum_{i=1}^{n} \frac{\partial^2 X}{\partial x_i^2} = (1 - yy')^{n/2+1} \sum_{i=1}^{n} \frac{\partial^2 Y}{\partial y_i^2}. \tag{3}$$

Before proving the last expression, let us first directly prove that the following $n + 1$ functions satisfy the Laplace equation:

$$\Phi(x) = (1 - 2ax' + aa'xx')^{1-n/2} \tag{4}$$

as well as

$$\Psi(x) = (1 - 2ax' + aa'xx')^{-n/2}[(1 - aa')(x-a) - (x-a)(x-a)'a]. \tag{5}$$

((5) is a vector, having altogether n component functions.) We shall first prove that $\Phi(x)$ is a harmonic function:

$$\frac{\partial \Phi}{\partial x_i} = 2\left(1 - \frac{n}{2}\right)(1 - 2ax' + aa'xx')^{-n/2}(aa'x_i - a_i),$$

$$\sum_{i=1}^{n} \frac{\partial^2 \Phi}{\partial x_i^2} = 4\left(1 - \frac{n}{2}\right)\left(-\frac{n}{2}\right)(1 - 2ax' + aa'xx')^{-n/2-1} \sum_{i=1}^{n} (aa'x_i - a_i)^2$$

$$+ 2\left(1 - \frac{n}{2}\right)(1 - 2ax' + aa'xx')^{-n/2}naa'$$

$$= 0.$$

Next we must prove that $\Psi(x)$ is also a harmonic function:

$$\frac{\partial \Psi}{\partial x_i} = -n(1 - 2ax' + aa'xx')^{-n/2-1}(aa'x_i - a_i)$$

$$\times \left[(1 - aa')(x - a) - (x - a)(x - a)'a\right]$$

$$+ (1 - 2ax' + aa'xx')^{-n/2}\left[(1 - aa')e_i - 2(x_i - a_i)a\right],$$

where $e_i = (0, 0, \ldots, 0, 1, 0, \ldots, 0)$, where the ith component is 1 and all the rest are 0. Differentiating again,

$$\frac{\partial^2 \Psi}{\partial x_i^2} = n(n + 2)(1 - 2ax' + aa'xx')^{-n/2-2}$$

$$\times (aa'x_i - a_i)^2\left[(1 - aa')(x - a) - (x - a)(x - a)'a\right]$$

$$- n(1 - 2ax' + aa'xx')^{-n/2-1}aa'\left[(1 - aa')(x - a)\right.$$

$$\left. - (x - a)(x - a)'a\right] - 2n(1 - 2ax' + aa'xx')^{-n/2-1}$$

$$\times (aa'x_i - a_i)\left[(1 - aa')e_i - 2(x_i - a_i)a\right]$$

$$+ (1 - 2ax' + aa'xx')^{-n/2}(-2a).$$

Thus

$$\sum_{i=1}^{n} \frac{\partial^2 \Psi}{\partial x_i^2} = n(n + 2)(1 - 2ax' + aa'xx')^{-n/2-1}$$

$$\times aa'\left[(1 - aa')(x - a) - (x - a)(x - a)'a\right]$$

$$- n^2(1 - 2ax' + aa'xx')^{-n/2-1}aa'\left[(1 - aa')(x - a)\right.$$

$$\left. - (x - a)(x - a')a\right] - 2n(1 - 2ax' + aa'xx')^{-n/2-1}$$

$$\times \left[aa'(1 - aa')x - (1 - aa')a - 2aa'(xx' - ax)a\right.$$

$$\left. + 2(ax' - aa')a\right] - 2n(1 - 2ax' + aa'xx')^{-n/2}a = 0.$$

Now we are ready to prove (3): we must find the partial derivatives of

$$X = \Phi(x)Y(1 - aa')^{n/2-1}.$$

We have:

$$(1 - aa')^{1-n/2} \frac{\partial X}{\partial x_i} = \sum_{j=1}^{n} \frac{\partial Y}{\partial y_j} \frac{\partial y_j}{\partial x_i} \Phi(x) + \frac{\partial \Phi}{\partial x_i} Y,$$

$$(1 - aa')^{1-n/2} \frac{\partial^2 X}{\partial x_i^2} = \sum_{j=1}^{n} \sum_{k=1}^{n} \frac{\partial^2 Y}{\partial y_j \partial y_k} \frac{\partial y_j}{\partial x_i} \frac{\partial y_k}{\partial x_i} \Phi(x)$$

$$+ \sum_{j=1}^{n} \frac{\partial Y}{\partial y_j} \frac{\partial^2 y_j}{\partial x_i^2} \Phi(x)$$

$$+ 2 \sum_{j=1}^{n} \frac{\partial Y}{\partial y_j} \frac{\partial y_j}{\partial x_i} \frac{\partial \Phi}{\partial x_i} + \frac{\partial^2 \Phi}{\partial x_i^2} Y.$$

$$(1 - aa')^{1-n/2} \sum_{i=1}^{n} \frac{\partial^2 X}{\partial x_i^2} = \sum_{j=1}^{n} \sum_{k=1}^{n} \frac{\partial^2 Y}{\partial y_j \partial y_k} \sum_{i=1}^{n} \frac{\partial y_j}{\partial x_i} \frac{\partial y_k}{\partial x_i} \Phi(x) \qquad (6)$$

$$+ \sum_{j=1}^{n} \frac{\partial Y}{\partial y_j} \sum_{i=1}^{n} \left(\frac{\partial^2 y_j}{\partial x_i^2} \Phi(x) + 2 \frac{\partial y_j}{\partial x_i} \frac{\partial \Phi}{\partial x_i} \right)$$

$$+ \sum_{i=1}^{n} \frac{\partial^2 \Phi}{\partial x_i^2} Y$$

$$= \sum_{j=1}^{n} \sum_{k=1}^{n} \frac{\partial^2 Y}{\partial y_j \partial y_k} \sum_{i=1}^{n} \frac{\partial y_j}{\partial x_i} \frac{\partial y_k}{\partial x_i} \Phi(x)$$

$$+ \sum_{j=1}^{n} \frac{\partial Y}{\partial y_j} \left[\sum_{i=1}^{n} \frac{\partial^2 \Psi_j(x)}{\partial x_i^2} - \sum_{i=1}^{n} \frac{\partial^2 \Phi}{\partial x_i^2} y_j \right]$$

$$+ \sum_{i=1}^{n} \frac{\partial^2 \Phi}{\partial x_i^2} Y,$$

and the last two terms are equal to 0.

Again from

$$dy \, dy' = \frac{(1 - aa')^2}{(1 - 2ax' + aa'xx')^2} \, dx \, dx',$$

we obtain

$$\sum_{i=1}^{n} \frac{\partial y_i}{\partial x_j} \frac{\partial y_i}{\partial x_k} = \frac{(1 - aa')^2}{(1 - 2ax' + aa'xx')^2} \delta_{jk},$$

Multiplying by $\partial x_k / \partial y_i$ and summing over k gives

$$\frac{(1 - aa')^2}{(1 - 2ax' + aa'xx')^2} \frac{\partial x_j}{\partial y_l} = \sum_{k=1}^{n} \left(\sum_{i=1}^{n} \frac{\partial y_i}{\partial x_j} \frac{\partial y_i}{\partial x_k} \right) \frac{\partial x_k}{\partial y_l} = \frac{\partial y_l}{\partial x_j},$$

and thus

$$\sum_{i=1}^{n} \frac{\partial y_j}{\partial x_i} \frac{\partial y_k}{\partial x_i} = \sum_{i=1}^{n} \frac{\partial y_j}{\partial x_i} \frac{(1 - aa')^2}{(1 - 2ax' + aa'xx')^2} \frac{\partial x_i}{\partial y_k}$$

$$= \frac{(1 - aa')^2}{(1 - 2ax' + aa'xx')^2} \delta_{jk}.$$

From (6) we have

$$(1 - aa')^{1-n/2} \sum_{i=1}^{n} \frac{\partial^2 X}{\partial x_i^2} = \frac{(1 - aa')^2 \Phi(x)}{(1 - 2ax' + aa'xx')^2} \sum_{i=1}^{n} \frac{\partial^2 Y}{\partial y_i^2}$$

$$= (1 - aa')^2 (1 - 2ax' + aa'xx')^{-n/2-1} \sum_{i=1}^{n} \frac{\partial^2 Y}{\partial y_i^2}, \quad (7)$$

Using the relation

$$1 - xx' = (1 - 2ax' + aa'xx')(1 - yy')/(1 - aa'),$$

we get

$$(1 - xx')^{n/2+1} \sum_{i=1}^{n} \frac{\partial^2 X}{\partial x_i^2} = (1 - yy')^{n/2+1} \sum_{i=1}^{n} \frac{\partial^2 Y}{\partial y_i^2}$$

which is (3).

Remark.

(1) In reality, the equation

$$dw = \frac{dz}{(cz + d)^2},$$

is valid for all Möbius transformations

$$w = \frac{az + b}{cz + d}, \qquad ad - bc = 1.$$

Thus

$$dw\,d\bar{w} = \frac{dz\,d\bar{z}}{|cz + d|^4}.$$

One hopes to obtain general results along this line. Since the n-dimensional generalization of the real form of the Möbius transformation is a conformal transformation, we will be dealing with spherical geometry!

(2) The simplest form of the symmetry principle is the one which takes the harmonic functions in R_1 (fig. 1) (here we omit references to the conditions that must be imposed along the real axis) and extends them to R_2. Moreover, since a line and any circle are equivalent, in general then, all circular arcs will have the corresponding result. Generalizing to the n-dimensional case relative to any piece of a sphere, we may always apply the symmetry principle in order to do analytic continuation.

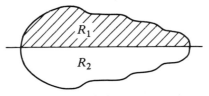

Figure 1.1

Since differential equation (5.1) is invariant under the inversion

$$y = x/xx',$$

the symmetry principle naturally may be generalized. We will use the result of of §6 of Chapter 3 to obtain an even more general result.

1.11 The Mean Value Formula for the Laplace Equation

Theorem 1. *If on the unit ball defined by* $xx' \leqslant 1$*, the function* $\Phi(x) = \Phi(\rho u)$ *satisfies the Laplace equation*

$$\sum_{i=1}^{n} \frac{\partial^2 \Phi}{\partial x_i^2} = \frac{1}{\rho^{n-1}} \frac{\partial}{\partial \rho} \left(\rho^{n-1} \frac{\partial \Phi}{\partial \rho} \right) + \frac{1}{\rho^2} \partial_u^2 \Phi = 0, \tag{1}$$

and has continuous second order partial derivatives, then, when $0 \leqslant \rho < 1$*,*

$$\frac{1}{\omega_{n-1}} \int \cdots \int_{uu' = 1} \Phi(\rho u) \dot{u} = \Phi(0). \tag{2}$$

PROOF. Let

$$F(\rho) = \frac{1}{\omega_{n-1}} \int \cdots \int_{uu' = 1} \Phi(\rho u) \dot{u}.$$

Differentiating under the integral, and then using (1), we obtain

$$\frac{1}{\rho^{n-3}} \frac{d}{d\rho} \left(\rho^{n-1} \frac{dF}{d\rho} \right) = \frac{1}{\omega_{n-1}} \int \cdots \int_{uu' = 1} \frac{1}{\rho^{n-3}} \frac{\partial}{\partial \rho} \left(\rho^{n-1} \frac{\partial \Phi}{\partial \rho} \right) \dot{u}$$

$$= -\frac{1}{\omega_{n-1}} \int \cdots \int_{uu' = 1} \partial_u^2 \Phi \dot{u}.$$

From

$$\partial_u^2 = \sum_{p=1}^{n-1} \left(\frac{1}{\sin^2 \theta_1 \cdots \sin^2 \theta_{p-1}} \frac{\partial^2}{\partial \theta_p^2} + (n-p-1) \frac{\operatorname{ctg} \theta_p}{\sin^2 \theta_1 \cdots \sin^2 \theta_{p-1}} \frac{\partial}{\partial \theta_p} \right)$$

$$= \sum_{p=1}^{n-1} \frac{1}{\sin^2 \theta_1 \cdots \sin^2 \theta_{p-1}} \cdot \frac{1}{\sin^{n-p-1} \theta_p} \frac{\partial}{\partial \theta_p} \left(\sin^{n-p-1} \theta_p \frac{\partial}{\partial \theta_p} \right)$$

and

$$\dot{u} = \sin^{n-2} \theta_1 \sin^{n-3} \theta_2 \cdots \sin^{n-p-1} \theta_p \cdots \sin \theta_{n-2} d\theta_1 \cdots d\theta_{n-1},$$

we have

$$\int \cdots \int_{uu'=1} \partial_u^2 \Phi(\rho u)\dot{u} = \sum_{p=1}^{n-1} J_p,$$

where

$$J_p = \int \cdots \int_{uu'=1} \frac{\partial}{\partial \theta_p} \left(\sin^{n-p-1} \theta_p \frac{\partial \Phi}{\partial \theta_p} \right) \frac{\sin^{n-2} \theta_1 \cdots \sin \theta_{n-2} \, d\theta_1 \cdots d\theta_{n-1}}{\sin^2 \theta_1 \cdots \sin^2 \theta_{p-1} \sin^{n-p-1} \theta_p}.$$

J_p is an $(n-1)$-fold integral, wherein the pth integral $(p < n - 1)$ is

$$\int_0^\pi \frac{\partial}{\partial \theta_p} \left(\sin^{n-p-1} \theta_p \frac{\partial \Phi}{\partial \theta_p} \right) d\theta_p = \sin^{n-p-1} \theta_p \frac{\partial \Phi}{\partial \theta_p} \Big|_0^\pi = 0.$$

Therefore, when $p < n - 1$, $J_p = 0$, and when $p = n - 1$, the $(n-1)$th integral of J_p is

$$\int_0^{2\pi} \frac{\partial^2 \Phi}{\partial \theta_{n-1}^2} d\theta_{n-1} = \frac{\partial \Phi}{\partial \theta_{n-1}} \Big|_0^{2\pi} = 0,$$

where we have made use of the periodicity of Φ with respect to θ_{n-1}. Thus also $J_{n-1} = 0$. In conclusion,

$$\int \cdots \int_{uu'=1} \partial_u^2 \Phi \dot{u} = 0.$$

Therefore

$$\frac{1}{\rho^{n-3}} \frac{d}{d\rho} \left(\rho^{n-1} \frac{dF}{d\rho} \right) = 0,$$

so that we deduce therefrom

$$\rho^{n-1} \frac{dF}{d\rho} = k,$$

k being a constant. When $\rho = 0$, we see that $k = 0$, and thus F is a constant. Again letting $\rho = 0$, we obtain formula (2). ☐

1.12 The Poisson Formula for the Laplace Equation

Returning to §10, let

$$Y(y) = \left(\frac{1 - 2ax' + aa'xx'}{1 - aa'} \right)^{n/2 - 1} X(x).$$

If $Y(y)$ satisfies the Laplace equation, then $X(x)$ does also. Applying the mean value formula to $Y(y)$, we get

$$\frac{1}{\omega_{n-1}} \int \cdots \int_{vv'=1} Y(v)\dot{v} = Y(0).$$

Since

$$Y(0) = \left(\frac{1 - 2aa' + (aa')^2}{1 - aa'}\right)^{n/2-1} X(a) = (1 - aa')^{n/2-1} X(a),$$

$$Y(v) = \left(\frac{1 - 2au' + aa'}{1 - aa'}\right)^{n/2-1} X(u)$$

and

$$\dot{v} = \left(\frac{1 - aa'}{1 - 2au' + aa'}\right)^{n-1} \dot{u},$$

we obtain:

$$(1 - aa')^{n/2-1} X(a) = Y(0) = \frac{1}{\omega_{n-1}} \int \cdots \int_{vv'=1} Y(v)\dot{v}$$

$$= \frac{1}{\omega_{n-1}} \int \cdots \int_{uu'=1} X(u) \left(\frac{1 - 2au' + aa'}{1 - aa'}\right)^{n/2-1}$$

$$\times \left(\frac{1 - aa'}{1 - 2au' + aa'}\right)^{n-1} \dot{u}.$$

The result is the Poisson formula for the Laplace equation:

$$X(a) = \frac{1}{\omega_{n-1}} \int \cdots \int_{uu'=1} \frac{1 - aa'}{(1 - 2au' + aa')^{n/2}} X(u)\dot{u}. \tag{1}$$

This gives the uniqueness of the solution to the Dirichlet problem for the Laplace equation,

1.13 A Brief Summary

Beginning with the unit disc, we may make a direct generalization to the unit ball

$$xx' < 1, \tag{1}$$

the boundary of which is

$$uu' = 1. \tag{2}$$

The volume element of the unit sphere is denoted by $\dot{u}$ and the total volume is

$$\int \cdots \int_{uu'=1} \dot{u} = \omega_{n-1}.$$

In this chapter the following points were made:
(a) Use the real form of the unit circle as the starting point.

The transformation group: the one generated by the transformation

$$y = \frac{(1 - aa')(x - a) - (x - a)(x - a)'a}{1 - 2ax' + aa'xx'}, \qquad aa' < 1 \qquad (3)$$

which takes $x = a$ into $y = 0$, and by the transformation

$$y = x\Gamma, \qquad \Gamma\Gamma' = I \qquad (4)$$

which leaves invariant the point 0. From this we have

$$1 - yy' = \frac{(1 - aa')(1 - xx')}{1 - 2ax' + aa'xx'}. \qquad (5)$$

This shows that the group takes (1) into itself. The differential invariant is:

$$\frac{dy\, dy'}{(1 - yy')^2} = \frac{dx\, dx'}{(1 - xx')^2}. \qquad (6)$$

The invariant second order partial differential operator is:

$$(1 - yy')^n \sum_{i=1}^{n} \frac{\partial}{\partial y_i}\left((1 - yy')^{2-n} \frac{\partial u}{\partial y_i}\right) = (1 - xx')^n \sum_{i=1}^{n} \frac{\partial}{\partial x_i}\left((1 - xx')^{2-n} \frac{\partial u}{\partial x_i}\right).$$

Thus we have the second order partial differential equation

$$(1 - xx')^n \sum_{i=1}^{n} \frac{\partial}{\partial x_i}\left((1 - xx')^{2-n} \frac{\partial u}{\partial x_i}\right) = 0. \qquad (7)$$

Transformations (3) and (4) also leave (2) invariant. The transformation rule of the volume element on (2) is

$$\dot{v} = \left(\frac{1 - aa'}{1 - 2au' + aa'}\right)^{n-1} \dot{u}, \qquad (8)$$

whence we derive the Poisson kernel. This, in turn, gives us the Poisson formula

$$\Phi(x) = \frac{1}{\omega_{n-1}} \int \cdots \int_{uu'=1} \left(\frac{1 - xx'}{1 - 2xu' + xx'}\right)^{n-1} \Phi(u)\dot{u}, \qquad (9)$$

which may be used to solve the boundary value problem (the Dirichlet problem) of (7).

(b) Use the real projective group as a starting point.

Transformation group: the one generated by the transformation

$$y = \frac{\sqrt{1 - aa'}(x - a)(I + \lambda a'a)}{1 - ax'},$$
$$\qquad (3')$$
$$\lambda = \frac{1 - \sqrt{1 - aa'}}{aa'\sqrt{1 - aa'}}$$

which takes $x = a$ into $y = 0$, and by the transformation

$$y = x\Gamma, \qquad \Gamma\Gamma' = I \tag{4'}$$

which leaves the point 0 unchanged. From this we have

$$1 - yy' = \frac{(1 - aa')(1 - xx')}{(1 - ax')^2}. \tag{5'}$$

The differential invariant is

$$\frac{dy(I - y'y)^{-1} dy'}{1 - yy'} = \frac{dx(I - x'x)^{-1} dx'}{1 - xx'}. \tag{6'}$$

The equation obtained from the invariant second order partial differential operator is

$$\sum_{i=1}^{n} \frac{\partial^2 \Phi}{\partial x_i^2} - \sum_{i,j=1}^{n} x_i x_j \frac{\partial^2 \Phi}{\partial x_i \partial x_j} - 2 \sum_{i=1}^{n} x_i \frac{\partial \Phi}{\partial x_i} = 0. \tag{7'}$$

The transformation rule of the volume element on (2) is

$$\dot{v} = \frac{(1 - aa')^{(n-1)/2}}{(1 - au')^{n-1}} \dot{u}. \tag{8'}$$

Thus the Poisson formula which solves the boundary value problem of (7') is

$$\Phi(x) = \frac{1}{\omega_{n-1}} \int \cdots \int_{uu'=1} \frac{(1 - xx')^{(n-1)/2}}{(1 - xu')^{n-1}} \Phi(u)\dot{u}. \tag{9'}$$

(c) The geometry discussed in (a) will serve as the introduction to the study of conformal mappings of spherical geometry in Chapter 3. Furthermore, the geometry discussed in (b) will serve as an introduction to partial differential equations of mixed types. The quadratic form of (7') is

$$I - x'x.$$

When $xx' < 1$, this square matrix is positive, definite; when $xx' > 1$, it is non-positive, or one positive and $(u - 1)$ negative in signature. Since the transformation of (2) is linear, it is therefore not invariant under inversion.

(d) The Laplacian is not invariant under (3); however, if we allow the functions on which it operates to also undergo a transformation, then we may still derive a covariance formula: subjecting x, y to transformation (3) gives

$$Y = \left(\frac{1 - 2ax' + aa'xx'}{1 - aa'}\right)^{n/2 - 1} X, \tag{10}$$

whence

$$(1 - xx')^{n/2 + 1} \sum_{i=1}^{n} \frac{\partial^2 X}{\partial x_i^2} = (1 - yy')^{n/2 + 1} \sum_{i=1}^{n} \frac{\partial^2 Y}{\partial y_i^2}. \tag{7''}$$

In §12 the Poisson formula for the Laplace equation was derived

$$X(x) = \frac{1}{\omega_{n-1}} \int \cdots \int_{uu'=1} \frac{1 - xx'}{(1 - 2xu' + xx')^{n/2}} X(u)\dot{u}.$$

(e) The generalization from the unit circle to the unit sphere is only the first step of an extensive development. The following book of the author gives part of this information: L. K. Hua, *Harmonic Analysis of Functions of Several Complex Variables in the Classical Domains*, (tr. by L. Ebner and A. Korányi) Amer. Math. Soc., Providence, 1963.

CHAPTER 2
Fourier Analysis and the Expansion Formulas for Harmonic Functions

2.1 A Few Properties of Spherical Functions[1]

For the ease of understanding, let us begin by explaining a few properties of spherical polynomials:

When $\lambda > -\frac{1}{2}$, the *spherical polynomial* $P_m^{(\lambda)}$ is defined by

$$P_m^{(\lambda)}(\xi) = \sum_{0 \leqslant l \leqslant m/2} (-1)^l \frac{\Gamma(m-l+\lambda)}{\Gamma(\lambda)l!(m-2l)!} (2\xi)^{m-2l} \tag{1}$$

it is an mth degree polynomial, and at times we even define $P_{-1}^{(\lambda)}(\xi) = 0$. It is not difficult to calculate that

$$P_0^{(\lambda)}(\xi) = 1, \qquad P_1^{(\lambda)}(\xi) = 2\lambda\xi,$$

$$P_2^{(\lambda)}(\xi) = 2\lambda(\lambda+1)\xi^2 - \lambda,$$

$$P_3^{(\lambda)}(\xi) = (4/3)\lambda(\lambda+1)(\lambda+2)\xi^3 - 2\lambda(\lambda+1)\xi, \ldots .$$

In general, (1) may be deduced from the recursion formula

$$mP_m^{(\lambda)}(\xi) = 2(m+\lambda-1)\xi P_{m-1}^{(\lambda)}(\xi) - (m+2\lambda-2)P_{m-2}^{(\lambda)}(\xi) \tag{2}$$

[1] Translator's note: In the original the author uses the term *hyperspherical functions*.

32

as follows: the right hand side of (2) is equal to

$$(m + \lambda - 1) \sum_{0 \leqslant l \leqslant (m-1)/2} (-1)^l \frac{\Gamma(m - 1 - l + \lambda)}{(\lambda)l!(m - 1 - 2l)!} (2\xi)^{m - 2l}$$

$$-(m + 2\lambda - 2) \sum_{0 \leqslant l \leqslant (m-2)/2} (-1)^l \frac{\Gamma(m - 2 - l + \lambda)}{\Gamma(\lambda)l!(m - 2 - 2l)!} (2\xi)^{m - 2l - 2}$$

$$= \sum_{0 \leqslant l \leqslant m/2} (-1)^l \left[\frac{\Gamma(m - 1 - l + \lambda)(m + \lambda - 1)}{\Gamma(\lambda)l!(m - 1 - 2l)!} \right.$$

$$\left. - \frac{\Gamma(m - 1 - l + \lambda)(m + 2\lambda - 2)}{\Gamma(\lambda)(l - 1)!(m - 2l)!} \right] (2\xi)^{m - 2l}$$

$$= \sum_{0 \leqslant l \leqslant m/2} (-1)^l \frac{\Gamma(m - 1 - l + \lambda)}{\Gamma(\lambda)l!(m - 2l)!}$$

$$\times \left[(m + \lambda - 1)(m - 2l) - (m + 2\lambda - 2)l \right] (2\xi)^{m - 2l}$$

$$= m \sum_{0 \leqslant l \leqslant m/2} (-1)^l \frac{\Gamma(m - l + \lambda)}{\Gamma(\lambda)l!(m - 2l)!} (2\xi)^{m - 2l}. \qquad \square$$

From recursion formula (2) we use induction to immediately obtain the result

$$\sum_{m=0}^{n} (\lambda + m)P_m^{(\lambda)}(\xi) = \frac{1}{2} \frac{(n + 2\lambda)P_n^{(\lambda)}(\xi) - (n + 1)P_{n+1}^{(\lambda)}(\xi)}{1 - \xi}. \qquad (3)$$

Multiplying (2) by ρ^{m-1} and summing over m gives

$$\sum_{m=0}^{\infty} m\rho^{m-1}P_m^{(\lambda)}(\xi) = 2 \sum_{m=0}^{\infty} (m + \lambda - 1)\xi P_{m-1}^{(\lambda)}(\xi)\rho^{m-1}$$

$$- \sum_{m=0}^{\infty} (m + 2\lambda - 2)P_{m-2}^{(\lambda)}(\xi)\rho^{m-1}.$$

Let

$$h(\rho) = \sum_{m=0}^{\infty} P_m^{(\lambda)}(\xi)\rho^m,$$

then this expansion may be rewritten as

$$h'(\rho) = 2\xi\rho^{1-\lambda}[\rho^\lambda h(\rho)]' - \rho^{2-2\lambda}[\rho^{2\lambda}h(\rho)]'$$
$$= 2\xi[\lambda h(\rho) + \rho h'(\rho)] - [2\lambda\rho h(\rho) + \rho^2 h'(\rho)],$$

or

$$h'(\rho)/h(\rho) = 2\lambda(\xi - \rho)/(1 - 2\xi\rho + \rho^2).$$

Then, using $h(0) = P_0^{(\lambda)}(\xi) = 1$, and integrating the resulting expression, we immediately get

$$h(\rho) = (1 - 2\xi\rho + \rho^2)^{-\lambda}.$$

Thus we have obtained the *generating function*:

$$(1 - 2\xi\rho + \rho^2)^{-\lambda} = \sum_{m=0}^{\infty} P_m^{(\lambda)}(\xi)\rho^m. \tag{4}$$

Differentiating formula (1) gives

$$\frac{d}{d\xi} P_m^{(\lambda)}(\xi) = 2 \sum_{0 \le l \le (m-1)/2} (-1)^l \frac{\Gamma(m - l + \lambda)}{\Gamma(\lambda)l!(m - 2l - 1)!} (2\xi)^{m-1-2l}$$

$$= 2\lambda \sum_{0 \le l \le (m-1)/2} (-1)^l \frac{\Gamma(m - 1 - l + \lambda + 1)}{\Gamma(\lambda + 1)l!(m - 1 - 2l)!} (2\xi)^{m-1-2l}.$$

Therefore we obtain the differential recursion formula:

$$(d/d\xi)P_m^{(\lambda)}(\xi) = 2\lambda P_{m-1}^{(\lambda+1)}(\xi). \tag{5}$$

It is also fairly simple to prove:

$$(1 - \xi^2)\frac{d^2}{d\xi^2} P_m^{(\lambda)}(\xi) - (2\lambda + 1)\xi \frac{d}{d\xi} P_m^{(\lambda)}(\xi) + m(m + 2\lambda)P_m^{(\lambda)}(\xi) = 0.$$

If we let

$$\eta = (1 - \xi^2)^{\lambda - 1/2} P_m^{(\lambda)}(\xi), \tag{6}$$

then η satisfies the differential equation

$$(1 - \xi^2)\frac{d^2\eta}{d\xi^2} + (2\lambda - 3)\xi \frac{d\eta}{d\xi} + (m + 1)(m + 2\lambda - 1)\eta = 0. \tag{7}$$

Now we prove the *Rodrique formula*:

$$(1 - \xi^2)^{\lambda - 1/2} P_m^{(\lambda)}(\xi) = \frac{(-2)^m}{m!} \frac{\Gamma(m + \lambda)\Gamma(m + 2\lambda)}{\Gamma(\lambda)\Gamma(2m + 2\lambda)} \left(\frac{d}{d\xi}\right)^m (1 - \xi^2)^{m+\lambda - 1/2}. \tag{8}$$

Before proving this formula, let us first consider the identity gotten from (1):

$$mP_m^{(\lambda)}(\xi) = (m + 2\lambda - 1)\xi P_{m-1}^{(\lambda)}(\xi) - 2\lambda(1 - \xi^2)P_{m-2}^{(\lambda+1)}(\xi), \tag{9}$$

Again by induction, the right side of the preceding expression equals

$$(m + 2\lambda - 1)\xi(1 - \xi^2)^{-\lambda + 1/2} \frac{(-2)^{m-1}}{(m-1)!} \frac{\Gamma(m - 1 + \lambda)}{\Gamma(\lambda)}$$

$$\times \frac{\Gamma(m - 1 + 2\lambda)}{\Gamma(2m + 2\lambda - 2)} \left(\frac{d}{d\xi}\right)^{m-1} (1 - \xi^2)^{m + \lambda - 3/2}$$

$$- 2\lambda(1 - \xi^2)^{-\lambda + 1/2} \frac{(-2)^{m-2}}{(m-2)!} \frac{\Gamma(m - 1 + \lambda)}{\Gamma(\lambda + 1)}$$

$$\times \frac{\Gamma(m + 2\lambda)}{\Gamma(2m + 2\lambda - 2)} \left(\frac{d}{d\xi}\right)^{m-2} (1 - \xi^2)^{m + \lambda - 3/2}$$

$$= \frac{(-2)^{m-1} \Gamma(m - 1 + \lambda)\Gamma(m + 2\lambda)}{(m-1)!\Gamma(\lambda)\Gamma(2m + 2\lambda - 2)} (1 - \xi^2)^{-\lambda + 1/2}$$

$$\times \left[\xi \left(\frac{d}{d\xi}\right)^{m-1} (1 - \xi^2)^{m + \lambda - 3/2} + (m - 1)\left(\frac{d}{d\xi}\right)^{m-2} (1 - \xi^2)^{m + \lambda - 3/2}\right]$$

$$= \frac{(-2)^{m-1} \Gamma(m - 1 + \lambda)\Gamma(m + 2\lambda)}{(m-1)!\Gamma(\lambda)\Gamma(2m + 2\lambda - 2)} (1 - \xi^2)^{-\lambda + 1/2}$$

$$\times \left(\frac{d}{d\xi}\right)^{m-1} \{(1 - \xi^2)^{m + \lambda - 3/2}\xi\} \quad \text{(from the Leibnitz formula)}$$

$$= \frac{(-2)^{m-1} \Gamma(m - 1 + \lambda)\Gamma(m + 2\lambda)}{(m-1)!\Gamma(\lambda)\Gamma(2m + 2\lambda - 2)} \left(\frac{-1}{2m + 2\lambda - 1}\right)$$

$$\times (1 - \xi^2)^{-\lambda + 1/2} \left(\frac{d}{d\xi}\right)^{m} (1 - \xi^2)^{m + \lambda - 1/2}$$

$$= \frac{-(-2)^{m-1} \Gamma(m - 1 + \lambda)\Gamma(m + 2\lambda)(2m + 2\lambda - 2)}{(m-1)!\Gamma(\lambda)\Gamma(2m + 2\lambda - 2) \cdot (2m + 2\lambda - 2)(2m + 2\lambda - 1)}$$

$$\times (1 - \xi^2)^{-\lambda + 1/2} \left(\frac{d}{d\xi}\right)^{m} (1 - \xi^2)^{m + \lambda - 1/2}$$

$$= \frac{(-2)^{m} \Gamma(m + \lambda)\Gamma(m + 2\lambda)}{(m-1)!\Gamma(\lambda)\Gamma(2m + 2\lambda)} (1 - \xi^2)^{-\lambda + 1/2} \left(\frac{d}{d\xi}\right)^{m} (1 - \xi^2)^{m + \lambda - 1/2}.$$

2.2 Orthogonality Properties

Suppose $f(\xi)$ is a function on the interval $[-1, +1]$ having m continuous derivatives. From the Rodrique formula we know that

$$\int_{-1}^{1} f(\xi) P_m^{(\lambda)}(\xi)(1 - \xi^2)^{\lambda - 1/2} \, d\xi$$

$$= \frac{(-2)^{m} \Gamma(m + \lambda)\Gamma(m + 2\lambda)}{m!\Gamma(\lambda)\Gamma(2m + 2\lambda)} \int_{-1}^{1} f(\xi) \left(\frac{d}{d\xi}\right)^{m} (1 - \xi^2)^{m + \lambda - 1/2} \, d\xi. \quad (1)$$

Using integration by parts,

$$\int_{-1}^{1} f(\xi)\left(\frac{d}{d\xi}\right)^{m}(1-\xi^2)^{m+\lambda-1/2}\,d\xi = f(\xi)\left(\frac{d}{d\xi}\right)^{m-1}(1-\xi^2)^{m+\lambda-1/2}\Big|_{-1}^{1}$$

$$-\int_{-1}^{1} f'(\xi)\left(\frac{d}{d\xi}\right)^{m-1}(1-\xi^2)^{m+\lambda-1/2}\,d\xi.$$

Now, since $\lambda > -1/2$, we have

$$\left(\frac{d}{d\xi}\right)^{m-1}(1-\xi^2)^{m+\lambda-1/2}\Big|_{-1}^{1} = 0,$$

whence

$$\int_{-1}^{1} f(\xi)\left(\frac{d}{d\xi}\right)^{m}(1-\xi^2)^{m+\lambda-1/2}\,d\xi = -\int_{-1}^{1} f'(\xi)\left(\frac{d}{d\xi}\right)^{m-1}(1-\xi^2)^{m+\lambda-1/2}\,d\xi.$$

Continuing in this manner, we finally obtain

$$\int_{-1}^{1} f(\xi)p_m^{(\lambda)}(\xi)(1-\xi^2)^{\lambda-1/2}\,d\xi$$

$$= \frac{2^m \Gamma(m+\lambda)\Gamma(m+2\lambda)}{m!\,\Gamma(\lambda)\Gamma(2m+2\lambda)}\int_{-1}^{1}(1-\xi^2)^{m+\lambda-1/2}\left(\frac{d}{d\xi}\right)^{m}f(\xi)\,d\xi. \quad (2)$$

If $f(\xi)$ is an mth degree polynomial, and the coefficient of its highest term equals a, then

$$\int_{-1}^{1} f(\xi)P_m^{(\lambda)}(\xi)(1-\xi^2)^{\lambda-1/2}\,d\xi = \frac{2^m \Gamma(m+\lambda)\Gamma(m+2\lambda)}{\Gamma(\lambda)\Gamma(2m+2\lambda)}$$

$$\times a\int_{-1}^{1}(1-\xi^2)^{m+\lambda-1/2}\,d\xi$$

$$= \frac{2^m \Gamma(m+\lambda)\Gamma(m+2\lambda)\Gamma(m+\lambda+\frac{1}{2})\Gamma(\frac{1}{2})}{\Gamma(\lambda)\Gamma(2m+2\lambda)\Gamma(m+\lambda+1)}\,a$$

$$= \frac{2^{-m-2\lambda+1}\pi\Gamma(m+2\lambda)}{\Gamma(\lambda)\Gamma(m+\lambda+1)}\,a, \quad (3)$$

where we have used the identity

$$\Gamma(x)\Gamma(x+1/2) = 2^{1-2x}\pi^{1/2}\Gamma(2x).$$

In particular, if we take $f(\xi) = P_l^{(\lambda)}(\xi)$, then we have from (1.1)

$$a = \begin{cases} 0, & l < m, \\[2mm] 2^m\,\dfrac{\Gamma(m+\lambda)}{\Gamma(\lambda)m!}, & l = m, \end{cases}$$

and then

$$\int_{-1}^{1} P_l^{(\lambda)}(\xi) P_m^{(\lambda)}(\xi)(1 - \xi^2)^{\lambda - 1/2} \, d\xi$$

$$= \begin{cases} 0, & l \neq m, \\ \dfrac{2^{1-2\lambda}\pi\Gamma(m + 2\lambda)}{[\Gamma(\lambda)]^2(m + \lambda)\Gamma(m + 1)}, & l = m. \quad (4) \end{cases}$$

This is the orthogonality property of the spherical functions.

Again in (2), take $f(\xi) = \xi^l$, then when $l \geqslant m$,

$$\int_{-1}^{1} \xi^l P_m^{(\lambda)}(\xi)(1 - \xi^2)^{\lambda - 1/2} \, d\xi$$

$$= \binom{l}{m} \frac{2^m \Gamma(m + \lambda)\Gamma(m + 2\lambda)}{\Gamma(\lambda)\Gamma(2m + 2\lambda)} \int_{-1}^{1} \xi^{l-m}(1 - \xi^2)^{m + \lambda - 1/2} \, d\xi. \quad (5)$$

When $l - m$ is odd, this integral equals 0; if $l - m = 2k$ is even, then from

$$\int_{-1}^{1} \xi^{l-m}(1 - \xi^2)^{m + \lambda - 1/2} \, d\xi = \frac{\Gamma(k + \frac{1}{2})\Gamma(m + \lambda + \frac{1}{2})}{\Gamma(k + m + \lambda + 1)},$$

we obtain

$$\int_{-1}^{1} \xi^l P_m^{(\lambda)}(\xi)(1 - \xi^2)^{\lambda - 1/2} \, d\xi$$

$$\qquad\qquad\qquad\qquad \text{if } l < m \text{ or } l - m \text{ is odd}$$

$$= \begin{cases} 0, \\ \dfrac{\pi}{2^{1+2k-1}} \dfrac{l!}{k!(l - 2k)!} \dfrac{\Gamma(l - 2k + 2\lambda)}{\Gamma(\lambda)\Gamma(l - k + \lambda + 1)}, \\ \qquad\qquad\qquad\qquad \text{if } l - m = 2k. \quad (6) \end{cases}$$

Since $P_l^{(\lambda)}(\xi)$ is an lth degree polynomial, we see that any mth degree polynomial may be expressed as

$$f(\xi) = \sum_{l=0}^{m} a_l P_l^{(\lambda)}(\xi). \quad (7)$$

Multiplying by $P_l^{(\lambda)}(\xi)(1 - \xi^2)^{\lambda - 1/2}$ and integrating from -1 to 1 gives

$$a_l = 2^{2\lambda - 1} \frac{(\Gamma(\lambda))^2(l + \lambda)\Gamma(l + 1)}{\pi\Gamma(l + 2\lambda)} \int_{-1}^{1} f(\xi) P_l^{(\lambda)}(\xi)(1 - \xi^2)^{\lambda - 1/2} \, d\xi. \quad (8)$$

An immediate consequence of this is

Theorem 1. Let $f(\xi)$ be an mth degree polynomial with coefficient a_l $(0 \leqslant l \leqslant m - 1)$ as defined in (8). If all the a_l are equal to 0, then $f(\xi)$ and $P_m^{(\lambda)}(\xi)$ differ by a constant multiple.

Combining (6), (7) and (8) gives the *expansion formula*:

$$\xi^m = \frac{m!\,\Gamma(\lambda)}{2^m} \sum_{0 \leqslant k \leqslant m/2} \frac{m - 2k + \lambda}{k!\,\Gamma(m - k + \lambda + 1)} P^{(\lambda)}_{m-2k}(\xi). \tag{9}$$

From this we prove the identity:

$$P^{(v)}_m(\xi) = \frac{\Gamma(\lambda)}{\Gamma(v)} \sum_{0 \leqslant k \leqslant m/2} c_k P^{(\lambda)}_{m-2k}(\xi), \quad \text{for all } v > \lambda > -1/2, \tag{10}$$

where

$$c_k = \frac{m - 2k + \lambda}{k!} \cdot \frac{\Gamma(k + v - \lambda)}{\Gamma(v - \lambda)} \cdot \frac{\Gamma(m + v - k)}{\Gamma(m + \lambda + 1 - k)}. \tag{11}$$

Before giving the proof, first note that there is a finite difference formula related to the Γ function. Define $\Delta f(x) = f(x + 1) - f(x)$ to be the *1st order finite difference* of the function $f(x)$, and by induction define $\Delta^q f(x) = \Delta^{q-1}[\Delta f(x)]$ to be the *qth order finite difference* of $f(x)$. It is not difficult to prove

$$\Delta^q f(x) = \sum_{l=0}^{q} (-1)^l \binom{q}{l} f(x + q - l).$$

Since

$$\Delta \frac{\Gamma(\alpha + x)}{\Gamma(\beta + x)} = \frac{\Gamma(\alpha + x + 1)}{\Gamma(\beta + x + 1)} - \frac{\Gamma(\alpha + x)}{\Gamma(\beta + x)}$$

$$= (\alpha - \beta) \frac{\Gamma(\alpha + x)}{\Gamma(\beta + x + 1)},$$

we see that when $\alpha > \beta$

$$\Delta^q \frac{\Gamma(\alpha + x)}{\Gamma(\beta + x)} = \frac{\Gamma(\alpha - \beta + 1)}{\Gamma(\alpha - \beta - q + 1)} \frac{\Gamma(\alpha + x)}{\Gamma(\beta + x + q)}. \tag{12}$$

We now come to the proof of (11). From (1.1) and (9),

$$P^{(v)}_m(\xi) = \sum_{0 \leqslant s \leqslant m/2} (-1)^s \frac{\Gamma(v + m - s)}{\Gamma(v)\Gamma(s + 1)\Gamma(m - 2s + 1)} (2\xi)^{m-2s}$$

$$= \frac{\Gamma(\lambda)}{\Gamma(v)} \sum_{0 \leqslant s \leqslant m/2} (-1)^s \frac{\Gamma(v + m - s)}{\Gamma(s + 1)} \sum_{0 \leqslant k \leqslant m/2 - s}$$

$$\times \frac{m - 2s - 2k + \lambda}{k!\,\Gamma(m - 2s - k + \lambda + 1)} P^{(\lambda)}_{m-2k-2s}(\xi)$$

$$= \frac{\Gamma(\lambda)}{\Gamma(v)} \sum_{0 \leqslant t \leqslant m/2} c_t P^{(\lambda)}_{m-2t}(\xi),$$

where

$$c_t = \sum_{s+k=t} (-1)^s \frac{\Gamma(v+m-s)(m-2s-2k+\lambda)}{s!k!\Gamma(m-2s-k+\lambda+1)}$$

$$= \frac{m-2t+\lambda}{t!} \sum_{s=0}^{t} (-1)^s \binom{t}{s} \frac{\Gamma(v+m-s)}{\Gamma(m-t-s+\lambda+1)}$$

$$= \frac{(m-2t+\lambda)\Gamma(t+v-\lambda)\Gamma(v+m-t)}{t!\Gamma(v-\lambda)\Gamma(m-t+\lambda+1)}. \qquad \square$$

2.3 The Boundary Value Problem

We must again begin with the unit circle: if there exists a Fourier series on the circumference, i.e.

$$f(e^{i\theta}) \sim a_0/2 + \sum_{n=1}^{\infty} (a_n \cos n\theta + b_n \sin n\theta), \qquad (1)$$

$$\begin{cases} a_0 = \frac{1}{\pi} \int_0^{2\pi} f(e^{i\theta})\, d\theta, \quad a_n = \frac{1}{\pi} \int_0^{2\pi} f(e^{i\theta}) \cos n\theta\, d\theta, \\ b_n = \frac{1}{\pi} \int_0^{2\pi} f(e^{i\theta}) \sin n\theta\, d\theta, \end{cases} \qquad (2)$$

then the function

$$f(\rho e^{i\theta}) = a_0/2 + \sum_{n=1}^{\infty} (a_n \cos n\theta + b_n \sin n\theta)\rho^n \qquad (3)$$

is the harmonic function which takes (1) as the boundary value. This point is very easily seen. First of all, $\rho^n \cos n\theta$, $\rho^n \sin n\theta$ are the harmonic functions taking $\cos n\theta$, $\sin n\theta$ as boundary values. Since the Laplace equation is linear, the claim follows.

Our goal now is to generalize this idea to the sphere. However, we must first point out that substitution of (2) into (3) gives

$$f(\rho e^{i\theta}) = \frac{1}{2\pi} \int_0^{2\pi} f(e^{i\psi})\, d\psi + \sum_{n=1}^{\infty} \frac{\rho^n}{\pi} \int_0^{2\pi} f(e^{i\psi})$$

$$\times (\cos n\psi \cos n\theta + \sin n\psi \sin n\theta)\, d\psi \qquad (4)$$

$$= \frac{1}{2\pi} \int_0^{2\pi} f(e^{i\psi})\, d\psi + \sum_{n=1}^{\infty} \frac{\rho^n}{\pi} \int_0^{2\pi} f(e^{i\psi}) \cos n(\theta-\psi)\, d\psi$$

$$= \frac{1}{2\pi} \int_0^{2\pi} f(e^{i\psi}) \left(1 + 2\sum_{n=1}^{\infty} \rho^n \cos n(\theta-\psi)\right) d\psi$$

$$= \frac{1}{2\pi} \int_0^{2\pi} f(e^{i\psi}) \frac{1-\rho^2}{1-2\rho\cos(\theta-\psi)+\rho^2}\, d\psi, \qquad (5)$$

from which the Poisson formula again follows.

(4) suggests the possibility of expressing any harmonic function $f(\rho u)$ in the unit ball as

$$f(\rho u) = \sum_{l=0}^{\infty} \rho^l \frac{1}{\omega_{n-1}} \int \cdots \int_{vv'=1} f(v) \Phi_l(u, v) \dot{v}. \tag{6}$$

If this were possible, then we could hope that $f(\rho u)$ would be the harmonic function taking

$$\sum_{l=0}^{\infty} \frac{1}{\omega_{n-1}} \int \cdots \int_{vv'=1} f(v) \Phi_l(u, v) \dot{v} \tag{7}$$

as boundary value. In reality, we ought to have begun with the harmonic analysis of the unit ball, first obtaining a complete orthonormal system of functions on the sphere, and then study (6). However, this course of action would have taken more time, and furthermore would have required a definite knowledge of group representations. We now take an opposite route, but one which is more convenient: first expand the Poisson kernel before doing anything else.

The Poisson kernel of the Laplace equation has the following expression: from (1.4), (2.10) and (2.11) we know that when $x = \rho v$, $vv' = 1$,

$$\frac{1 - xx'}{(1 - 2xu' + xx')^{n/2}} = (1 - \rho^2) \sum_{m=0}^{\infty} P_m^{(n/2)}(uv') \rho^m$$

$$= (1 - \rho^2) \frac{\Gamma\left(\dfrac{n}{2} - 1\right)}{\Gamma\left(\dfrac{n}{2}\right)} \sum_{m=0}^{\infty} \sum_{0 \leqslant k \leqslant m/2} c_k P_{m-2k}^{(n/2-1)}(uv') \rho^m,$$

where $c_k = m - 2k + 1/2n - 1$. Letting $l = m - 2k$ gives

$$\frac{1 - xx'}{(1 - 2xu' + xx')^{n/2}} = (1 - \rho^2) \sum_{l=0}^{\infty} \frac{l + \frac{1}{2}n - 1}{\frac{1}{2}n - 1} P_l^{(n/2-1)}(uv') \rho^l \sum_{k=0}^{\infty} \rho^{2k}$$

$$= \sum_{l=0}^{\infty} \frac{2l + n - 2}{n - 2} \rho^l P_l^{(n/2-1)}(uv').$$

Thus the Poisson integral formula may be written as

$$\frac{1}{\omega_{n-1}} \int \cdots \int_{vv'=1} \frac{(1 - \rho^2) f(v)}{(1 - 2\rho \cos uv' + \rho^2)^{n/2}} \dot{v}$$

$$= \frac{1}{\omega_{n-1}} \sum_{l=0}^{\infty} \frac{2l + n - 2}{n - 2} \rho^l \int \cdots \int_{vv'=1} P_l^{(n/2-1)}(uv') f(v) \dot{v}.$$

The above expression suggests that if a function $f(v)$ on a sphere has an expansion

$$f(u) \sim \frac{1}{\omega_{n-1}} \sum_{l=0}^{\infty} \frac{2l+n-2}{n-2} \int \cdots \int_{vv'=1} P_l^{(n/2-1)}(uv')f(v)\dot{v},$$ (8)

then we hope that there exists a harmonic function

$$f(\rho u) \sim \frac{1}{\omega_{n-1}} \sum_{l=0}^{\infty} \frac{2l+n-2}{n-2} \rho^l \int \cdots \int_{vv'=1} P_l^{(n/2-1)}(uv')f(v)\dot{v}$$ (9)

which takes $f(u)$ as boundary value.

Expansion (8) is called the *Laplace series*, and it is the natural generalization of the *Fourier series*. The spherical functions $\{P_l^{(n/2-1)}(\xi)\}$ are also called the *Legendre functions* or the *Legendre polynomials*.

We will not delve deeply into the convergence problem related to (8) (for further information, consult Chen Jian-Gong, *Theory of Orthogonal Functions*, Science Press, Beijing (in Chinese).), but we can at least point out that the convergence of (8) guarantees the convergence (9). On the other hand, the convergence of (9) does not imply the convergence of (8). As a matter of fact, it does not even imply that (8) defines a function on the boundary.

In the same way that Theorem 1 was proved in §1.7, so may the following be proved:

If $f(v)$ is a continuous function, then

$$\lim_{\rho \to 1} \frac{1}{\omega_{n-1}} \sum_{l=0}^{\infty} \frac{2l+n-2}{n-2} \rho^l \int \cdots \int_{vv'=1} P_l^{(n/2-1)}(uv')f(v)\dot{v} = f(u).$$ (10)

This suggests the existence of "Abel's theorem" and "Tauberian theorems", that is to say, if

$$\lim_{N \to \infty} \frac{1}{\omega_{n-1}} \sum_{l=0}^{N} \frac{2l+n-2}{n-2} \int \cdots \int_{vv'=1} P_l^{(n/2-1)}(uv')f(v)\dot{v} = s_0,$$ (11)

then it is perhaps true that

$$\lim_{\rho \to 1} \frac{1}{\omega_{n-1}} \sum_{l=0}^{\infty} \frac{2l+n-2}{n-2} \rho^l \int \cdots \int_{vv'=1} P_l^{(n/2-1)}(uv')f(v)\dot{v} = s_0.$$ (12)

This is just the usual *Abel's theorem* for power series, which is of course true.

On the other hand, if (12) holds and moreover

$$\int \cdots \int_{vv'=1} P_l^{(n/2-1)}(uv')f(v)\dot{v} = O\left(\frac{1}{l^2}\right),$$

then (11) also holds.

2.4 Generalized Functions on the Sphere

Now let us consider a method of defining a generalized function on the sphere. Given a continuous function on the sphere, we have a harmonic function in the unit ball which takes it as boundary value. On the other hand, a function which is everywhere harmonic in the unit ball does not necessarily have a continuous boundary value; in fact it may not even approach a function on the boundary. However we may abstractly define: a *generalized function on the sphere* is the boundary value of a harmonic function inside the ball.

Concretely speaking, suppose we have an expansion

$$f(\rho u) = \frac{1}{\omega_{n-1}} \sum_{l=0}^{\infty} \frac{2l + n - 2}{n - 2} \rho^l \int \cdots \int_{vv' = 1} P_l^{(n/2 - 1)}(uv') f(v) \dot{v}, \tag{1}$$

and, moreover, suppose

$$\lim_{l \to \infty} \left| \int \cdots \int_{vv' = 1} P_l^{(n/2 - 1)}(uv') f(v) \dot{v} \right|^{-1/l} \leqslant 1,$$

then series (1) converges in the unit ball and defines a harmonic function there. We now define a *generalized function* on the unit sphere as a formal Laplace series

$$\frac{1}{\omega_{n-1}} \sum_{l=0}^{\infty} \frac{2l + n - 2}{n - 2} \int \cdots \int_{vv' = 1} P_l^{(n/2 - 1)}(uv') f(v) \dot{v}.$$

Thus by limiting our discussion to the circle, the scope of the generalized functions so defined is much broader than that of L. Schwarz, not to mention that it is much easier to work with.

2.5 Harmonic Analysis on the Sphere

The Laplace series is not the most precise generalization of the Fourier series, for it was not based on the consideration of a complete, orthogonal system of functions.

Let γ represent the unit sphere of an n-dimensional space, i.e. the set of all vectors $u = (u_1, \ldots, u_n)$ satisfying

$$uu' = 1. \tag{1}$$

In spherical coordinates, the sphere γ is expressed as

$$
\begin{aligned}
u_1 &= \cos\theta_1, \\
u_2 &= \sin\theta_1 \cos\theta_2, \\
&\;\;\vdots \\
u_{n-1} &= \sin\theta_1 \sin\theta_2 \cdots \sin\theta_{n-2} \cos\theta_{n-1}, \\
u_n &= \sin\theta_1 \sin\theta_2 \cdots \sin\theta_{n-2} \sin\theta_{n-1},
\end{aligned}
\tag{2}
$$

where

$$
0 \leqslant \theta_r \leqslant \pi \quad (1 \leqslant r \leqslant n-2), \qquad 0 \leqslant \theta_{n-1} \leqslant 2\pi.
\tag{3}
$$

The volume element of the sphere is

$$
\dot{u} = \sin^{n-2}\theta_1 \sin^{n-3}\theta_2 \cdots \sin\theta_{n-2} \, d\theta_1 \cdots d\theta_{n-1},
\tag{4}
$$

and the total volume equals

$$
\begin{aligned}
\omega = \omega_{n-1} &= \int \cdots \int_{uu'=1} \dot{u} \\
&= \int_0^\pi \sin^{n-2}\theta_1 \, d\theta_1 \cdots \int_0^\pi \sin\theta_{n-2} \, d\theta_{n-2} \int_0^{2\pi} d\theta_{n-1} \\
&= \frac{2\pi^{n/2}}{\Gamma(\tfrac{1}{2}n)}.
\end{aligned}
\tag{5}
$$

The main goal of harmonic analysis on the sphere is to find a system of functions

$$
\varphi_i(u) = \varphi_i(u_1, \ldots, u_n), \qquad i = 0, 1, 2, \ldots,
$$

which are orthonormal on γ, i.e.

$$
\frac{1}{\omega_{n-1}} \int \cdots \int_{uu'=1} \varphi_i(u)\varphi_j(u)\dot{u} = \delta_{ij},
$$

and furthermore, from "any function", $f(u)$ can be approximated by linear combinations of $\varphi_i(u)$, that is to say, given $\varepsilon > 0$, there exist $c_0, \ldots, c_n$, such that

$$
\left| f(u) - \sum_{i=0}^M c_i\varphi_i \right| < \varepsilon.
$$

Furthermore, define

$$
\sum_{i=0}^\infty c_i\varphi_i(u), \qquad c_i = \frac{1}{\omega_{n-1}} \int \cdots \int_{vv'=1} f(v)\varphi_i(v)\dot{v}
$$

to be the *Fourier series of the function* $f(u)$.

The best way to generalize Fourier series is to exhibit a concrete orthonormal system $\{\varphi_i\}$. However to do so would require more preparation (for details, see the author's book, *Harmonic Analysis of Functions of Several*

Variables in the Classical Domains, Chapter 7, §7.2). The Laplace series is merely the case of adding indiscriminately the components of a certain irreducible representation. That is to say, $P_m^{(n/2-1)}(uv')$ of the Laplace series is obtained by nothing other than adding certain $\varphi_i(u)\varphi_i(v)$ of

$$\sum_{i=0}^{\infty} c_i \varphi_i(u) = \frac{1}{\omega_{n-1}} \sum_{i=0}^{\infty} \int \cdots \int_{vv'=1} f(v)\varphi_i(u)\varphi_i(v)\dot{v}.$$

2.6 Expansion of the Poisson Kernel of Invariant Equations

Consider once again the expansion of

$$\left(\frac{1-xx'}{1-2xu'+xx'}\right)^{n-1}.$$

Let $x = \rho v$, $vv' = 1$, then (1.4), (2.10) and (2.11) give

$$\left(\frac{1-xx'}{1-2xu'+xx'}\right)^{n-1} = \left(\frac{1-\rho^2}{1-2\rho uv'+\rho^2}\right)^{n-1}$$

$$= (1-\rho^2)^{n-1} \sum_{m=0}^{\infty} P_m^{(n-1)}(uv')\rho^m$$

$$= (1-\rho^2)^{n-1} \sum_{m=0}^{\infty} \frac{\Gamma(\tfrac{1}{2}n-1)}{\Gamma(n-1)} \sum_{0 \leqslant k \leqslant m/2} c_k P_{m-2k}^{(n/2-1)}(uv')\rho^m$$

$$= (1-\rho^2)^{n-1} \frac{\Gamma(\tfrac{1}{2}n-1)}{\Gamma(n-1)} \sum_{l=0}^{\infty} \psi_l(\rho) P_l^{(n/2-1)}(uv'), \tag{1}$$

where

$$\psi_l(\rho) = \rho^l \sum_{k=0}^{\infty} c_k \rho^{2k}$$

$$= \rho^l \sum_{k=0}^{\infty} \frac{l + \tfrac{1}{2}n - 1}{k!} \frac{\Gamma(k+\tfrac{1}{2}n)}{\Gamma(\tfrac{1}{2}n)} \frac{\Gamma(l+k+n-1)}{\Gamma(l+k+\tfrac{1}{2}n)} \rho^{2k}$$

$$= \rho^l \frac{\Gamma(l+n-1)}{\Gamma(l+\tfrac{1}{2}n-1)} F(\tfrac{1}{2}n, l+n-1; l+\tfrac{1}{2}n; \rho^2),$$

and $F(\alpha, \beta, \gamma; x)$ is the hypergeometric series.

From the properties of a hypergeometric series:

$$F(\alpha, \beta, \gamma; x) = (1-x)^{\gamma-\alpha-\beta} F(\gamma-\gamma, \alpha-\beta, \gamma; x),$$

and when $\alpha + \beta - \gamma < 0$,

$$F(\alpha, \beta; \gamma; 1) = \frac{\Gamma(\alpha - \alpha - \beta)\Gamma(\gamma)}{\Gamma(\alpha - \alpha)\Gamma(\gamma - \beta)},$$

which in turn gives

$$\left(\frac{1 - \rho^2}{1 - 2\rho uv' + \rho^2}\right)^{n-1} = \frac{\Gamma(\tfrac{1}{2}n - 1)}{\Gamma(n-1)} \sum_{l=0}^{\infty} \frac{\Gamma(l + n - 1)}{\Gamma(l + \tfrac{1}{2}n - 1)}$$

$$\rho^l F(l, -\tfrac{1}{2}n + 1; l + \tfrac{1}{2}n; \rho^2) P_l^{(n/2-1)}(uv')$$

$$= \frac{\Gamma(\tfrac{1}{2}n - 1)}{\Gamma(n-1)} \sum_{l=0}^{\infty} \frac{\Gamma(l + n - 1)\Gamma(n - 1)\Gamma(l + \tfrac{1}{2}n)}{\Gamma(l + \tfrac{1}{2}n - 1)\Gamma(\tfrac{1}{2}n) \ (l + n - 1)}$$

$$\times \ \tau_l(\rho)\rho^l P_l^{(n/2-1)}(uv')$$

$$= \sum_{l=0}^{\infty} \frac{2l + n - 2}{n - 2} \tau_l(\rho)\rho^l P_l^{(n/2-1)}(uv'), \tag{2}$$

where

$$\tau_l(\rho) = \frac{\Gamma(\tfrac{1}{2}n)\Gamma(l + n - 1)}{\Gamma(n - 1)\Gamma(l + \tfrac{1}{2}n)} F(l, -\tfrac{1}{2}n + 1; l + \tfrac{1}{2}n; \rho^2), \tag{3}$$

satisfying

$$\lim_{\rho \to 1} \tau_l(\rho) = 1. \tag{4}$$

Thus the Poisson formula may be rewritten as

$$\frac{1}{\omega_{n-1}} \int \cdots \int_{vv'=1} \left(\frac{1 - \rho^2}{1 - 2\rho uv' + \rho^2}\right)^{n-1} f(v)\dot{v}$$

$$= \frac{1}{\omega_{n-1}} \sum_{l=0}^{\infty} \frac{2l + n - 2}{n - 2} \tau_l(\rho)\rho^l \int \cdots \int_{vv'=1} P_l^{(n/2-1)}(uv')f(v)\dot{v}. \tag{5}$$

In other words, if a given function $f(v)$ has a converging Laplace series

$$\frac{1}{\omega_{n-1}} \sum_{l=0}^{\infty} \frac{2l + n - 2}{n - 2} \int \cdots \int_{vv'=1} P_l^{(n/2-1)}(uv')f(v)\dot{v}, \tag{6}$$

then in the unit ball the invariant equation has solution

$$\frac{1}{\omega_{n-1}} \sum_{l=0}^{\infty} \frac{2l + n - 2}{n - 2} \tau_l(\rho) \cdot \rho^l \int \cdots \int_{vv'=1} P_l^{(n/2-1)}(uv')f(v)\dot{v},$$

and this solution takes $f(v)$ as its boundary value.

Remark. The way to proceed is to hope that we may preserve intact the previous harmonic analysis of the sphere, and by going so far as rewriting

ρ^l as $\rho^l \tau_l(\rho)$. An alternate way is to preserve the multiple ρ^l but instead change the corresponding Legendre function. For example, rewrite (1) as

$$(1 - \rho^2)^{n-1} \sum_{m=0}^{\infty} P_m^{(n-1)}(uv')\rho^m = \sum_{l=0}^{\infty} Q_l(uv')\rho^l,$$

where

$$Q_l(\xi) = \sum_{0 \le k \le \min(n-1,\, l/2)} (-1)^k \binom{n-1}{k} P_{l-2k}^{(n-1)}(\xi).$$

Then continue on as before.

2.7 Completeness

We have previously proved that for any continuous function $\varphi(u)$ on the sphere, we always have

$$\varphi(u) = \lim_{\rho \to 1} \frac{1}{\omega_{n-1}} \sum_{l=0}^{\infty} \frac{2l - n + 2}{n - 2} \rho^l \int \cdots \int_{vv'=1} P_l^{(n/2-1)}(uv')\varphi(v)\dot{v}.$$

Whence, given any $\varepsilon > 0$, there exists ρ such that

$$\left| \varphi(u) - \frac{1}{\omega_{n-1}} \sum_{l=0}^{\infty} \frac{2l - n + 2}{n - 2} \rho^l \int \cdots \int_{vv'=1} P_l^{(n/2-1)}(uv')\varphi(v)\dot{v} \right| < \frac{1}{2}\varepsilon,$$

and there exists N such that

$$\left| \varphi(u) - \frac{1}{\omega_{n-1}} \sum_{l=0}^{N} \frac{2l - n + 2}{n - 2} \rho^l \int \cdots \int_{vv'=1} P_l^{(n/2-1)}(uv')\varphi(v)\dot{v} \right| < \varepsilon. \tag{1}$$

We shall now prove the following: if a continuous function $\varphi(u)$ on the sphere satisfies

$$\int \cdots \int_{uu'=1} \varphi(u) P_l^{(n/2-1)}(uv')\dot{u} = 0, \qquad l = 0, 1, 2, \ldots, \tag{2}$$

then $\varphi(u) \equiv 0$.

This is an extremely simple fact to prove, since we know from (1) that

$$\frac{1}{\omega_{n-1}} \int \cdots \int_{uu'=1} \left(\varphi(u) - \frac{1}{\omega_{n-1}} \sum_{l=0}^{N} \frac{2l - n + 2}{n - 2} \rho^l \right.$$

$$\left. \times \int \cdots \int_{vv'=1} P_l^{(n/2-1)}(uv')\varphi(v)\dot{v} \right)^2 \dot{u} < \varepsilon. \tag{3}$$

Multiplying out the integrand, we know from (2) that the left side equals

$$\frac{1}{\omega_{n-1}} \int \cdots \int_{uu'=1} [\varphi(u)]^2 \dot{u} + \frac{1}{\omega_{n-1}} \int \cdots \int_{uu'=1} \left(\frac{1}{\omega_{n-1}} \sum_{l=0}^{N} \frac{2l-n+2}{n-2} \rho^l \right.$$

$$\left. \times \int \cdots \int_{vv'=1} P_l^{(n/2-1)}(uv')\varphi(v)\dot{v} \right)^2 \dot{u} \geqslant \frac{1}{\omega_{n-1}} \int \cdots \int_{uu'=1} [\varphi(u)]^2 \dot{u}.$$

Then from (3), given any $\varepsilon > 0$, we always have

$$\frac{1}{\omega_{n-1}} \int \cdots \int_{uu'=1} [\varphi(u)]^2 \dot{u} < \varepsilon,$$

which is impossible unless $\varphi(u) \equiv 0$. $\qquad\square$

2.8 Solving the Partial Differential Equation $\partial_u^2 \Phi = \lambda \Phi$

Consider on the sphere the partial differential equation (cf. equation (6.5) of Chapter 1):

$$\partial_u^2 \Phi = \lambda \Phi. \tag{1}$$

If this equation has a solution for some λ, then this λ is called an *eigenvalue*. Let us now prove:

Theorem 1. *The only eigenvalues are $\lambda = -l(l+n-2)$ ($l = 0, 1, 2, \ldots$), and $P_l^{(n/2-1)}(uv')$ is the solution corresponding to such a λ.*

PROOF. We claim:

$$\partial_u^2 P_l^{(n/2-1)}(uv') = -l(l+n-2)P_l^{(n/2-1)}(uv'). \tag{2}$$

To this end, choose $v = (1, 0, \ldots, 0)$, then (2) simplifies to

$$\left(\frac{\partial^2}{\partial \theta_1^2} + (n-2) \operatorname{ctg} \theta_1 \frac{\partial}{\partial \theta_1} \right) P_l^{(n/2-1)}(\cos \theta_1)$$
$$= -l(l+n-2)P_l^{(n/2-1)}(\cos \theta_1). \tag{3}$$

Change variables by letting $\cos \theta_1 = \xi$; then (3) is equivalent to

$$\left[(1-\xi^2) \frac{\partial^2}{\partial \xi^2} - (n-1)\xi \frac{\partial}{\partial \xi} \right] P_l^{(n/2-1)}(\xi) = -l(l+n-2)P_l^{(n/2-1)}(\xi),$$

which is just formula (1.6). Thus when $v = (1, 0, \ldots, 0)$, (2) holds.

Suppose v is any unit vector, then there exists an orthogonal matrix Γ such that $v\Gamma = (1, 0, \ldots, 0)$. Now by introducing spherical coordinates,

$$u\Gamma = (\cos\theta_1,\ \sin\theta_1 \cos\theta_2, \ldots),$$

the problem then simplifies to exactly the same situation as above. So (2) is proved.

We further claim: if

$$\partial_u^2 \Phi = \lambda\Phi, \qquad \partial_u^2 \Psi = \mu\Psi, \qquad \lambda \neq \mu, \tag{4}$$

then

$$\int \cdots \int_{vv'=1} \Phi(v)\Psi(v)\dot{v} = 0. \tag{5}$$

Indeed,

$$\int \cdots \int_{vv'=1} \Phi(v)\partial_v^2 \Psi(v)\dot{v} = \sum_{p=1}^{n-1} J_p,$$

where

$$J_p = \int \cdots \int_{vv'=1} \Phi(v)\frac{\partial}{\partial\theta_p}\left(\sin^{n-p-1}\theta_p \frac{\partial\Psi}{\partial\theta_p}\right) \frac{\sin^{n-2}\theta_1 \cdots \sin\theta_{n-2}\,d\theta_1 \cdots d\theta_{n-1}}{\sin^2\theta_1 \cdots \sin^2\theta_{p-1}\sin^{n-p-1}\theta_p}.$$

When $p < n - 1$, J_p contains the simple integral

$$\int_0^\pi \Phi \frac{\partial}{\partial\theta_p}\left(\sin^{n-p-1}\theta_p \frac{\partial\Psi}{\partial\theta_p}\right) d\theta_p$$

$$= \Phi \sin^{n-p-1}\theta_p \frac{\partial\Psi}{\partial\theta_p}\bigg|_0^\pi - \int_0^\pi \frac{\partial\Phi}{\partial\theta_p}\sin^{n-p-1}\theta_p \frac{\partial\Psi}{\partial\theta_p}\,d\theta_p$$

$$= \int_0^\pi \Psi \frac{\partial}{\partial\theta_p}\left(\sin^{n-p-1}\theta_p \frac{\partial\Phi}{\partial\theta_p}\right) d\theta_p.$$

Thus Φ and Ψ have been interchanged. A similar situation results for the case when $p = n - 1$, but here we must make use of the periodicity of the functions Φ and Ψ with respect to θ_{n-1}. Therefore we have

$$\int \cdots \int_{vv'=1} \Phi(v)\partial_v^2 \Psi(v)\dot{v} = \int \cdots \int_{vv'=1} (v)\partial_v^2 \Phi(v)\dot{v}. \tag{6}$$

This, together with (4), give

$$(\lambda - \mu) \int \cdots \int_{vv'=1} \Phi(v)\Psi(v)\dot{v} = 0,$$

which is just expression (5).

It remains to prove that other than

$$\lambda = -l(l + n - 2) \qquad (l = 0, 1, 2, \ldots),$$

there are no other eigenvalues for (1). Simultaneously we shall prove that when $\lambda = -l(l + n - 2)$, $P_l^{(n/2-1)}(uv')$ is the unique corresponding eigenfunction. (Note: this is more than just giving the eigenfunctions, since given any point v on the sphere, there exists a $P_l^{(n/2-1)}(uv')$, and these eigenfunctions constitute a linear space. The dimension of this space is

$$\binom{n+l-1}{l} - \binom{n+l-3}{l-2},$$

a fact which we will not prove now.)

Suppose λ is another eigenvalue and $\Phi(v)$ is the corresponding eigenfunction, then

$$\int \cdots \int_{vv'=1} \Phi(v) P_l^{(n/2-1)}(uv')\dot{v} = 0, \qquad l = 0, 1, 2, \ldots .$$

By completeness, not only is $\Phi(v) = 0$, but it is also seen that when $\lambda = -l(l + n - 2)$, $P_l^{(n/2-1)}(uv')$ represents all of the eigenfunctions corresponding to this eigenvalue. $\qquad\square$

2.9 Remarks

It is also possible to obtain all results for spherical functions from the properties of the Poisson kernel. For example:

(1) First of all, the function

$$H(x, y) = \frac{1 - xx'yy'}{(1 - 2xy' + xx'yy')^{n/2}}, \qquad xx' < 1, \ yy' < 1 \qquad (1)$$

is a harmonic function of $(x_1, \ldots, x_n)$.

PROOF.

$$\frac{\partial}{\partial x_i} H(x, y) = \frac{-2x_i yy'}{(1 - 2xy' + xx'yy')^{n/2}} + \frac{n(1 - xx'yy')(y_i - x_i yy')}{(1 - 2xy' + xx'yy')^{n/2+1}}$$

and

$$\frac{\partial^2}{\partial x_i^2} H(x, y) = -\frac{2yy'}{(1 - 2xy' + xx'yy')^{n/2}}$$
$$- \frac{4nx_i yy'(y_i - x_i yy')}{(1 - 2xy' + xx'yy')^{n/2+1}} - \frac{n(1 - xx'yy')yy'}{(1 - 2xy' + xx'yy')^{n/2}}$$
$$+ n(n + 2)\frac{(1 - xx'yy')(y_i - x_i yy')^2}{(1 - 2xy' + xx'yy')^{n/2+2}}.$$

Thus

$$\sum_{i=1}^{n} \frac{\partial^2}{\partial x_i^2} H(x, y) = -\frac{2nyy'}{(1 - 2xy' + xx'yy')^{n/2}}$$

$$-\frac{4nyy'(xy' - xx'yy')}{(1 - 2xy' + xx'yy')^{n/2+1}} - \frac{n^2(1 - xx'yy')yy'}{(1 - 2xy' + xx'yy')^{n/2+1}}$$

$$+\frac{n(n + 2)(1 - xx'yy')[yy' - 2xy'yy' + xx'(yy')^2]}{(1 - 2xy' + xx'yy')^{n/2+2}}$$

$$= 0. \qquad\qquad \square$$

(2) We have the equality

$$H(x, y) = \frac{1}{\omega_{n-1}} \int \cdots \int_{vv' = 1} H(x, v)H(v, y)\dot{v},$$

(2)

$$xx' < 1, \qquad yy' < 1.$$

The proof of this result is very simple, the reason being that $H(x, v)$ is just the Poisson kernel, while $H(v, y)$ is the boundary value of the harmonic function $H(x, y)$.

(3) Letting $x = \rho u$, $y = rv$, and letting

$$H(\rho u, rv) = \frac{1 - \rho^2 r^2}{(1 - 2\rho r uv' + \rho^2 r^2)^{n/2}}$$

$$= \sum_{l=0}^{\infty} Q_l(uv')(\rho r)^l$$

(3)

be the power series expansion, then substituting into (2) gives

$$\sum_{l=0}^{\infty} Q_l(uv')(\rho r)^l = \frac{1}{\omega_{n-1}} \int \cdots \int_{ww' = 1} \sum_{p=0}^{\infty} Q_p(uw')\rho^p \sum_{q=0}^{\infty} Q_q(wv')r^q \dot{w}.$$

Now integrating term by term and comparing coefficients, we have

$$\frac{1}{\omega_{n-1}} \int \cdots \int_{ww' = 1} Q_p(uw')Q_q(wv')\dot{w} = \begin{cases} 0, & p \neq q, \\ Q_p(uv'), & p = q. \end{cases}$$

(4)

(4) Taking the special case $u = v = (1, 0, \ldots, 0)$, $w = (\cos\theta_1, \sin\theta_1 \cos\theta_2, \ldots)$ and $\dot{w} = \sin^{n-2}\theta_1 \cdots \sin\theta_{n-2} d\theta_1 \cdots d\theta_{n-1}$, we have

$$\frac{1}{\omega_{n-1}} \int_0^\pi Q_p(\cos\theta_1)Q_q(\cos\theta_1) \sin^{n-2}\theta_1 \, d\theta$$

$$\times \int_0^\pi \sin^{n-3}\theta_2 \, d\theta_2 \cdots \int_0^\pi \sin\theta_{n-2} \, d\theta_{n-2} \int_0^{2\pi} d\theta_{n-1} = \begin{cases} 0, & p \neq q \\ Q_p(1), & p = q. \end{cases}$$

Now note that

$$\omega_{n-1} = \frac{2\pi^{n/2}}{\Gamma\left(\dfrac{n}{2}\right)}, \qquad \int_0^\pi \sin^p\theta\, d\theta = \sqrt{\pi}\,\frac{\Gamma\left(\dfrac{p+1}{2}\right)}{\Gamma\left(\dfrac{p}{2}+1\right)}$$

and that $Q_p(1)$ is the coefficient of x^p in the expansion of $(1+x)(1-x)^{-(n-1)}$, i.e.

$$Q_p(1) = \frac{(n-1)n\cdots(n+p-3)(n+2p-2)}{p!}$$

$$= \frac{(n+p-3)!(n+2p-2)}{(n-2)!p!}.$$

Hence, we have

$$\int_0^\pi Q_p(\cos\theta_1)Q_q(\cos\theta_1)\sin^{n-2}\theta_1\,d\theta_1$$

$$= \pi^{1/2}\,\frac{\Gamma\left(\dfrac{n-1}{2}\right)}{\Gamma\left(\dfrac{n}{2}\right)}\,\frac{(n+p-3)!}{(n-2)!p!}\,(n+2p-2)\,\delta_{pq}.$$

Letting $\cos\theta_1 = \xi$ yields

$$\int_{-1}^1 Q_p(\xi)Q_q(\xi)(1-\xi^2)^{(n-3)/2}\,d\xi = \frac{(n+2p-2)2^{2-n}\pi\cdot(n+p-3)!}{\Gamma\left(\dfrac{n}{2}\right)^2 p!}\,\delta_{pq}. \quad (5)$$

(5) Substituting the power series

$$H(\rho u, v) = \sum_{l=0}^\infty Q_l(uv')\rho^l$$

into the Laplace equation, we obtain

$$\frac{1}{\rho^{n-1}}\frac{\partial}{\partial\rho}\left(\rho^{n-1}\frac{\partial}{\partial\rho}(H(\rho u, v))\right) + \frac{1}{\rho^2}\partial_u^2 H(\rho u, v) = 0,$$

and comparing the coefficients of ρ^{l-2} gives

$$l(n+l-2)Q_l(uv') + \partial_u^2 Q_l(uv') = 0.$$

Choosing the special case $v = (1, 0, \ldots, 0)$ and $\cos\theta = \xi$, we then have that $Q_1(\xi)$ satisfies the second order differential equation:

$$(1-\xi^2)\frac{d^2 Q_l}{d\xi^2} - (n-1)\xi\frac{dQ_l}{d\xi} + l(l+n-2)Q_l = 0. \quad (6)$$

(6) Using similar methods as those above, we may prove the completeness of $Q_l(uv')$ on the sphere and the completeness, as well as other properties, of Q_l on $(-1, +1)$, etc.

The upshot of all this is that frequent use of the Poisson formula would yield all the important properties of the spherical functions. This method of derivation would further unify the studies of spherical functions and the Laplace equation.

Exercise: Prove

$$\int_0^\pi Q_m(\xi\xi' + \sqrt{1 - \xi^2}\sqrt{1 - \xi'^2} \cos \theta) \sin^{n-3} \theta \, d\theta = cQ_m(\xi)Q_m(\xi'),$$

and determine c.

CHAPTER 3
Extended Space and Spherical Geometry

3.1 Quadratic Forms and Generalized Space

So far we have generalized from the unit disc to the unit ball. Now we shall take the whole plane (the Gaussian plane together with the point at infinity) and the Möbius group which acts on the plane and generalize the whole set-up to n-dimensional space. Our present discussion will be somewhat abstract, but the reader may draw an analogy with Chapter 1 or think of expressions of the transformations which leave the unit ball invariant in order to come to grips with this generalization.

We shall begin with a quadratic form,

$$x_1^2 + \cdots + x_n^2 - y_1 y_2 = 0, \tag{1}$$

whose matrix is the $(n + 2) \times (n + 2)$ square matrix

$$J = \begin{pmatrix} I^{(n)} & 0 & 0 \\ 0 & 0 & -\frac{1}{2} \\ 0 & -\frac{1}{2} & 0 \end{pmatrix}. \tag{2}$$

Let us consider the projective transformation

$$(\xi^*, \eta_1^*, \eta_2^*) = \rho(\xi, \eta_1, \eta_2)M, \tag{3}$$

where ξ, ξ^* are n-dimensional real vectors, η_1, η_2, η_1^*, η_2^* are real numbers, and M is such that

$$MJM' = J. \tag{4}$$

Obviously (3) keeps intact the relation

$$\xi\xi' = \eta_1\eta_2. \tag{5}$$

Let

$$M = \begin{pmatrix} T & u'_1 & u'_2 \\ v_1 & a & b \\ v_2 & c & d \end{pmatrix}, \tag{6}$$

then

$$\begin{cases} \xi^* = \rho(\xi T + \eta_1 v_1 + \eta_2 v_2), \\ \eta_1^* = \rho(\xi u'_1 + \eta_1 a + \eta_2 c), \\ \eta_2^* = \rho(\xi u'_2 + \eta_1 b + \eta_2 d). \end{cases} \tag{7}$$

Let us study the non-homogeneous coordinates: let $x = \xi/\eta_2$ (a vector) and $y = \xi^*/\eta_2^*$; from (5) we get

$$\frac{\eta_1}{\eta_2} = xx', \qquad \frac{\eta_1^*}{\eta_2^*} = yy'.$$

(7) becomes the transformation

$$y = \frac{xT + xx'v_1 + v_2}{xu'_2 + xx'b + d}, \tag{8}$$

$$yy' = \frac{xu'_1 + xx'a + c}{xu'_2 + xx'b + d}.$$

Note that the second equation in (8) may be derived from the first. From

$$M^{-1} = JM'J^{-1} = \begin{pmatrix} T' & -2v'_2 & -2v'_1 \\ -\frac{1}{2}u_2 & d & b \\ -\frac{1}{2}u_1 & c & a \end{pmatrix},$$

we see that the inverse transformation of (8) is

$$\begin{cases} x = \frac{yT' - \frac{1}{2}yy'u_2 - \frac{1}{2}u_1}{-2yv'_1 + yy'b + a}, \\ xx' = \frac{-2yv'_2 + yy'd + c}{-2yv'_1 + yy'b + a}. \end{cases} \tag{9}$$

All of the transformations of the form of (8) form a group; we call this group G.

The point corresponding to $\eta_2 = 0$ is called the *point at infinity*. Adding this point at infinity gives rise to a space, called the *extended space*. When $\eta_2 = 0$, then $\xi\xi' = 0$. Hence $\xi = 0$ and the point $(\xi, \eta_1, \eta_2) = \eta_1(0, 1, 0)$ is uniquely determined.

Note that

$$xu'_2 + xx'b + d = 0 \tag{10}$$

yields only one point. First, from the property of M^{-1}, we know that

$$-\tfrac{1}{4}u_2 u_2' - bd = 0.$$

If $b = 0$, then $u_2 = 0$, and (10) has no solutions, that is to say, (8) takes ∞ into ∞. If $b \neq 0$, then (10) may be rewritten as

$$\left(x - \frac{u_2}{b}\right)\left(x - \frac{u_2}{b}\right)' = 0,$$

i.e. (8) takes $x = u_2/b$ into ∞.

Problem: If a birational transformation takes the extended space into itself, is it necessarily of the form (8)?

It is not difficult to prove that the extended space is a transitive set; any two points may simultaneously be transformed into 0, ∞, so that the transformation leaving invariant 0 and ∞ is

$$y = \frac{1}{d}xT,$$

where $a = 1/d$, and T satisfies $TT' = I$. Thus any vector $x = \rho u$ may be transformed into e. Choosing $d = \rho$ and $uT = e$, then any three points may be transformed into any other three points. What is the invariant associated with any four points?

3.2 Differential Metric, Conformal Mappings

Differentiating

$$(y, yy', 1) = \rho(x, xx', 1)M \tag{1}$$

gives

$$(dy, 2y\,dy', 0) = [d\rho(x, xx', 1) + \rho(dx, 2x\,dx', 0)]M. \tag{2}$$

From

$$(dy, 2y\,dy', 0)J(dy, 2y\,dy', 0)' = dy\,dy'$$

and

$$(x, xx', 1)J(x, xx', 1)' = 0$$
$$(x, xx', 1)J(dx, 2x\,dx', 0)' = 0,$$

we obtain

$$dy\,dy' = \rho^2\,dx\,dx'. \tag{3}$$

Since

$$1 = \rho(xu_2' + xx'b + d),$$

therefore

$$dy\,dy' = \frac{dx\,dx'}{(xu_2' + xx'b + d)^2}. \tag{4}$$

That is to say, this transformation takes an infinitesimal round sphere into an infinitesimal round sphere, and is moreover conformal.

From the inverse transformation we obtain

$$dx\,dx' = \frac{dy\,dy'}{(-2yv'_1 + yy'b + a)^2}.\tag{5}$$

These two expressions suggest that

$$(xu'_2 + xx'b + d)(-2yv'_1 + yy'b + a) = 1.\tag{6}$$

A direct proof for (6) is very simple because from (1),

$$1 = \rho(xu'_2 + xx'b + d),$$

and from

$$(y, yy', 1)M^{-1} = \rho(x, xx', 1),$$

we have

$$\rho = -2yv'_1 + yy'b + a.$$

We shall now prove

$$\sum_{i=1}^{n}\frac{\partial^2 u}{\partial y_i^2} = \lambda^n \sum_{i=1}^{n}\frac{\partial}{\partial x_i}\left(\lambda^{2-n}\frac{\partial u}{\partial x_i}\right)$$

$$= \lambda^2 \sum_{i=1}^{n}\frac{\partial^2 u}{\partial x_i^2} + (2-n)\lambda \sum_{i=1}^{n}\frac{\partial\lambda}{\partial x_i}\frac{\partial u}{\partial x_i},\tag{7}$$

where $\lambda = xu'_2 + xx'b + d$.

PROOF. From expression (4),

$$\sum_{i=1}^{n}\frac{\partial y_i}{\partial x_s}\frac{\partial y_i}{\partial x_t} = \frac{1}{\lambda^2}\delta_{st}.\tag{8}$$

Then multiplying by $\partial x_t/\partial y_j$ and summing over t gives

$$\frac{\partial y_j}{\partial x_s} = \frac{1}{\lambda^2}\frac{\partial x_s}{\partial y_j}.\tag{9}$$

Thus

$$\frac{\partial u}{\partial y_i} = \sum_{j=1}^{n}\frac{\partial u}{\partial x_j}\frac{\partial x_j}{\partial y_i},$$

$$\sum_{i=1}^{n}\frac{\partial^2 u}{\partial y_i^2} = \sum_{j,k=1}^{n}\frac{\partial^2 u}{\partial x_j \partial x_k}\sum_{i=1}^{n}\frac{\partial x_j}{\partial y_i}\frac{\partial x_k}{\partial y_i} + \sum_{j=1}^{n}\frac{\partial u}{\partial x_j}\sum_{i=1}^{n}\frac{\partial^2 x_j}{\partial y_i^2}$$

$$= \lambda^4 \sum_{j,k=1}^{n}\frac{\partial^2 u}{\partial x_j \partial x_k}\sum_{i=1}^{n}\frac{\partial y_i}{\partial x_j}\cdot\frac{\partial y_j}{\partial x_k} + \sum_{j=1}^{n}\frac{\partial u}{\partial x_j}\sum_{i=1}^{n}\frac{\partial^2 x_j}{\partial y_i^2}$$

$$= \lambda^2 \sum_{i=1}^{n}\frac{\partial^2 u}{\partial x_i^2} + \sum_{j=1}^{n}\frac{\partial u}{\partial x_j}\sum_{i=1}^{n}\frac{\partial^2 x_j}{\partial y_i^2}.$$

Comparing this with (7), we see that it remains to prove:

$$\sum_{i=1}^{n} \frac{\partial^2 x_j}{\partial y_i^2} = (2-n)\lambda \frac{\partial \lambda}{\partial x_j}. \tag{10}$$

From (8) and (9) we have

$$\sum_{i=1}^{n} \frac{\partial x_i}{\partial y_p} \frac{\partial x_i}{\partial y_q} = \lambda^2 \delta_{pq}.$$

Differentiating,

$$\sum_{i=1}^{n} \frac{\partial^2 x_i}{\partial y_p^2} \frac{\partial x_j}{\partial y_q} + \sum_{i=1}^{n} \frac{\partial x_i}{\partial y_p} \frac{\partial^2 x_i}{\partial y_p \partial y_q} = \frac{\partial \lambda^2}{\partial y_p} \delta_{pq},$$

i.e.

$$\sum_{i=1}^{n} \frac{\partial^2 x_i}{\partial y_p^2} \frac{\partial x_i}{\partial y_q} + \frac{1}{2} \frac{\partial}{\partial y_q} \left(\sum_{i=1}^{n} \left(\frac{\partial x_i}{\partial y_p} \right)^2 \right) = \frac{\partial \lambda^2}{\partial y_p} \delta_{pq},$$

or

$$\sum_{i=1}^{n} \frac{\partial^2 x_i}{\partial y_p^2} \frac{\partial x_i}{\partial y_q} = \frac{\partial \lambda^2}{\partial y_p} \delta_{pq} - \frac{1}{2} \frac{\partial \lambda^2}{\partial y_q}.$$

Now multiplying by $\partial y_q / \partial x_r$ and summing over q, we obtain

$$\frac{\partial^2 x_r}{\partial y_p^2} = \frac{\partial \lambda^2}{\partial y_p} \frac{\partial y_p}{\partial x_r} - \frac{1}{2} \frac{\partial \lambda^2}{\partial x_r},$$

and summing over p gives (10) and thus (7). $\qquad\square$

Delving just a bit deeper into the problem, we utilize the identity

$$\frac{\partial}{\partial x_i} \left(r^2 \frac{\partial}{\partial x_i} (r^{-1}\Phi) \right) = r \frac{\partial^2 \Phi}{\partial x_i^2} - \Phi \frac{\partial^2 r}{\partial x_i^2}. \tag{11}$$

We can derive the following from (7): let

$$Y(y) = X(x)\lambda^{n/2-1},$$

then

$$\sum_{i=1}^{n} \frac{\partial^2 Y}{\partial y_i^2} = \lambda^n \sum_{i=1}^{n} \frac{\partial}{\partial x_i} \left((\lambda^{1-n/2})^2 \frac{\partial(\lambda^{n/2-1}X)}{\partial x_i} \right)$$

$$= \lambda^{n/2+1} \sum_{i=1}^{n} \frac{\partial^2 X}{\partial x_i^2} - \lambda^n X \sum_{i=1}^{n} \frac{\partial^2 \lambda^{n/2-1}}{\partial x_i^2}.$$

It is not difficult to prove directly that $\lambda^{n/2-1}$ is a harmonic function. Hence the second term equals 0, and we obtain

$$\sum_{i=1}^{n} \frac{\partial^2 Y}{\partial y_i^2} = \lambda^{n/2+1} \sum_{i=1}^{n} \frac{\partial^2 X}{\partial x_i^2}. \tag{12}$$

3.3 Mapping Spheres into Spheres

The sphere may be written as

$$\eta_1 yy' + y\xi' + \eta_2 = 0. \tag{1}$$

Symbolically, denote it by an $(n + 2)$-dimensional vector (ξ, η_1, η_2).
    Using transformation (1.8), the sphere (1) becomes

$$\eta_1(xu_1' + xx'a + c) + (xT + xx'v_1 + v_2)\xi' + \eta_2(xu_2' + xx'b + d) = 0.$$

This is also a sphere, to be denoted by $(\xi^*, \eta_1^*, \eta_2^*)$. So we have

$$\begin{pmatrix} \xi^{*\prime} \\ \eta_1^* \\ \eta_2^* \end{pmatrix} = \begin{pmatrix} T & u_1' & u_2' \\ v_1 & a & c \\ v_2 & b & d \end{pmatrix} \begin{pmatrix} \xi' \\ \eta_1 \\ \eta_2 \end{pmatrix},$$

i.e. we have

$$(\xi^*, \eta_1^*, \eta_2^*) = (\xi, \eta_1, \eta_2)M'. \tag{2}$$

By completing the square, (1) may be rewritten as

$$\left(y - \frac{\xi}{2\eta_1}\right)\left(y - \frac{\xi}{2\eta_1}\right)' = \frac{\xi\xi'}{4\eta_1^2} - \frac{\eta_2}{\eta_1} = \frac{\xi\xi' - 4\eta_1\eta_2}{4\eta_1^2}.$$

Therefore, depending on whether

$$(\xi, \eta_1, \eta_2)J^{-1}(\xi, \eta_1, \eta_2)' = \xi\xi' - 4\eta_1\eta_2 \gtreqless 0,$$

(1) represents respectively the real sphere, the point sphere or the imaginary sphere.
    It is not difficult to prove that transformation (1.8) transforms the real sphere, the point sphere and the imaginary sphere respectively into

$$xx' = 1, \qquad xx' = 0, \qquad xx' = -1.$$

The real sphere constitutes a transitive set. Note that the plane is a real sphere, and in particular, the set defined by $x_1 = 0$ is a real sphere.
    The transformation which takes the upper half-space $x_1 > 0$ into the unit ball $yy' < 1$ is

$$y = \frac{(xx' - 1, 2x_2, \ldots, 2x_n)}{(x_1 + 1)^2 + x_2^2 + \cdots + x_n^2}.$$

This fact may be easily proved as follows: we already have that

$$yy' = \frac{(xx' + 1)^2 + 4(xx' - x_1^2)}{(1 + 2x_1 + xx')^2} = \frac{(1 + xx')^2 - 4x_1^2}{(1 + xx' + 2x_1)^2}$$

$$= \frac{1 + xx' - 2x_1}{1 + xx' + 2x_1}$$

and

$$1 - yy' = \frac{4x_1}{1 + xx' + 2x_1},$$

so transform $x_1 > 0$ into $yy' < 1$. It also takes the plane $\xi_1 = 0$ into the unit sphere $vv' = 1$, i.e.

$$v = \frac{(\xi_2^2 + \cdots + \xi_n^2 - 1, 2\xi_2, \ldots, 2\xi_n)}{1 + \xi_2^2 + \cdots + \xi_n^2}.$$

Therefore,

$$1 - 2yv' + yy' = 2 \frac{1 + xx'}{1 + xx' + 2x_1}$$

$$- 2 \frac{(1 - xx')(1 - \xi_2^2 \cdots \xi_n^2) + 4(x_2\xi_2 + \cdots + x_n\xi_n)}{(1 + xx' + 2x_1)(1 + \xi_2^2 + \cdots + \xi_n^2)}$$

$$= 4 \frac{xx' + \xi_2^2 + \cdots + \xi_n^2 + 2(x_2\xi_2 + \cdots + x_n\xi_n)}{(1 + xx' + 2x_1)(1 + \xi_2^2 + \cdots + \xi_n^2)}$$

$$= 4 \frac{x_1^2 + (\xi_2 + x_2)^2 + \cdots + (\xi_n + x_n)^2}{(1 + xx' + 2x_1)(1 + \xi_2^2 + \cdots + \xi_n^2)}.$$

Consequently,

$$\left(\frac{1 - yy'}{1 - 2yv' + yy'}\right)^{n-1} \dot{v} = \left(\frac{x_1(1 + \xi_2^2 + \cdots + \xi_n^2)}{x_1^2 + (\xi_2 + x_2)^2 + \cdots + (\xi_n + x_n)^2}\right)^{n-1}$$

$$\times \frac{d\xi_2 \cdots d\xi_n}{[\frac{1}{2}(1 + \xi_2^2 + \cdots + \xi_n^2)]^{n-1}}$$

$$= \left(\frac{2x_1}{x_1^2 + (\xi_2 + x_2)^2 + \cdots + (\xi_n + x_n)^2}\right)^{n-1} d\xi_2 \cdots d\xi_n.$$

In other words, we have obtained the Poisson formula for the "upper-half" space:

$$\Phi(x_1, \ldots, x_n) = \frac{1}{\omega_{n-1}} \int_{-\infty}^{\infty} \cdots \int_{-\infty}^{\infty} \frac{(2x_1)^{n-1}\Phi(0, \xi_2, \ldots, \xi_n)\,d\xi_2 \cdots d\xi_n}{[x_1^2 + (\xi_2 + x_2)^2 + \cdots + (\xi_n + x_n)^2]^{n-1}},$$

and the partial differential equation it satisfies is

$$x_1^2 \sum_{i=1}^{n} \frac{\partial^2 \Phi}{\partial x_i^2} + (2 - n)x_1 \frac{\partial \Phi}{\partial x_1} = 0. \tag{3}$$

Furthermore the Poisson formula, which is derived from the Laplace equation related to the unit sphere, is

$$\Phi(y) = \frac{1}{\omega_{n-1}} \int \cdots \int_{vv'-1} \frac{1 - yy'}{(1 - 2yv' + yy')^{n/2}} \Phi(v)\dot{v}. \tag{4}$$

Define a new function $\Psi(x)$ by:

$$\Phi(y) = \Psi(x)\lambda^{n/2-1} = \Psi(x)(1 + xx' + 2x_1)^{n/2-1}. \tag{5}$$

Using

$$\Phi(v) = \Psi(0, \xi_2, \ldots, \xi_n)(1 + \xi_2^2 + \cdots + \xi_n^2)^{n/2-1},$$

we get

$$\frac{1 - yy'}{(1 - 2yv' + yy')^{n/2}} = \frac{x_1(1 + xx' + 2x_1)^{n/2-1}(1 + \xi_2^2 + \cdots + \xi_n^2)^{n/2}}{2^{n-2}(x_1^2 + (\xi_2 + x_2)^2 + \cdots + (\xi_n + x_n)^2)^{n/2}},$$

$$\dot{v} = \left(\frac{2}{1 + \xi_2^2 + \cdots + \xi_n^2}\right)^{n-1} d\xi_2 \cdots d\xi_n.$$

Thus from (4) and (5) we have

$$\Psi(x) = (1 + xx' + 2x_1)^{1-n/2}\Phi(y)$$

$$= (1 + xx' + 2x_1)^{1-n/2}\frac{1}{\omega_{n-1}}\int\cdots\int_{\omega'=1}\frac{1 - yy'}{(1 - 2yv' + yy')^{n/2}}\Phi(v)\dot{v}$$

$$= (1 + xx' + 2x_1)^{1-n/2}\frac{1}{\omega_{n-1}}$$

$$\cdot\int_{-\infty}^{\infty}\cdots\int\frac{x_1(1 + xx' + 2x_1)^{n/2-1}(1 + \xi_2^2 + \cdots + \xi_n^3)^{n/2}\Psi(0,\xi_2,\ldots,\xi_n)}{2^{n-2}(x_1^2 + (\xi_2 + x_2)^2 + \cdots + (\xi_n + x_n)^2)^{n/2}}$$

$$\cdot(1 + \xi_2^2 + \cdots + \xi_n^2)^{n/2-1}\left(\frac{2}{1 + \xi_2^2 + \cdots + \xi_n^2}\right)^{n-1}d\xi_2 \cdots d\xi_n$$

$$= \frac{1}{\omega_{n-1}}\int_{-\infty}^{\infty}\cdots\int\frac{2x_1\Psi(0,\xi_2,\ldots,\xi_n)}{(x_1^2 + (\xi_2 + x_2)^2 + \cdots + (\xi_n + x_n)^2)^{n/2}}d\xi_2 \cdots d\xi_n,$$

and this then is the Poisson formula for the Laplace equation on "upper-half" space.

3.4 Tangent Spheres and Chains of Spheres

What are the conditions for the tangency of two spheres

$$(\xi,\eta_1,\eta_2), (\xi^*,\eta_1^*,\eta_2^*)?$$

Naturally we must have

$$\left[\left(\frac{\xi}{2\eta_1} - \frac{\xi^*}{2\eta_1^*}\right)\left(\frac{\xi}{2\eta_1} - \frac{\xi^*}{2\eta_1^*}\right)'\right]^{1/2} = \sqrt{\frac{\xi\xi' - 4\eta_1\eta_2}{4\eta_1^2}} \pm \sqrt{\frac{\xi^*\xi^{*'} - 4\eta_1^*\eta_2^*}{4\eta_1^{*2}}},$$

i.e.

$$(2\eta_1\eta_2^* + 2\eta_2\eta_1^* - \xi\xi^{*'})^2 = (\xi\xi' - 4\eta_1\eta_2)(\xi^*\xi^{*'} - 4\eta_1^*\eta_2^*).$$

In other words, the determinant must satisfy

$$\left|\begin{pmatrix}\xi & \eta_1 & \eta_2 \\ \xi^* & \eta_1^* & \eta_2^*\end{pmatrix}J^{-1}\begin{pmatrix}\xi & \eta_1 & \eta_2 \\ \xi^* & \eta_1^* & \eta_2^*\end{pmatrix}'\right| = 0, \tag{1}$$

where J is as in (1.2). This suggests that we study the 2×2 matrix

$$\begin{pmatrix} \xi & \eta_1 & \eta_2 \\ \xi^* & \eta_1^* & \eta_2^* \end{pmatrix} J^{-1} \begin{pmatrix} \xi & \eta_1 & \eta_2 \\ \xi^* & \eta_1^* & \eta_2^* \end{pmatrix}' \equiv S. \tag{2}$$

We now define: the set consisting of spheres of the form

$$\lambda(\xi, \eta_1, \eta_2) + \mu(\xi^*, \eta_1^*, \eta_2^*)$$

is called a *chain of spheres*.

A chain of spheres may be represented by the $2 \times (n+2)$ matrix

$$X = \begin{pmatrix} \xi & \eta_1 & \eta_2 \\ \xi^* & \eta_1^* & \eta_2^* \end{pmatrix}.$$

If there is a 2×2 square matrix Q such that

$$QX = Y,$$

then X and Y represent the same chain of spheres.

Since the signature of J^{-1} has $(n+1)$ positive signs and one negative sign, S has consequently three possibilities: (i) positive definite, (ii) singular, but positive semi-definite, (iii) the signature has one positive sign and one negative sign. The corresponding chains of spheres are defined to be *elliptic*, *parabolic* and *hyperbolic*, respectively.

3.5 Orthogonal Spheres and Families of Spheres

If the angle between two spheres

$$(\xi, \eta_1 \eta_2), \quad (\xi^*, \eta_1^*, \eta_2^*)$$

is θ, then

$$\left(\frac{\xi}{2\eta_1} - \frac{\xi^*}{2\eta_1^*} \right) \left(\frac{\xi}{2\eta_1} - \frac{\xi^*}{2\eta_1^*} \right)' = \frac{\xi\xi' - 4\eta_1\eta_2}{4\eta_1^2} + \frac{\xi^*\xi^{*'} - 4\eta_1^*\eta_2^*}{4\eta_1^{*2}}$$

$$- 2 \sqrt{\frac{\xi\xi' - 4\eta_1\eta_2}{4\eta_1^2}} \sqrt{\frac{\xi^*\xi^{*'} - 4\eta_1^*\eta_2^*}{4\eta_1^{*2}}} \cos\theta,$$

that is to say,

$$\cos\theta = \frac{\xi\xi^{*'} - 2\eta_1\eta_2^* - 2\eta_2\eta_1^*}{\sqrt{(\xi\xi' - 4\eta_1\eta_2)(\xi^*\xi^{*'} - 4\eta_1^*\eta_2^*)}}$$

$$= \frac{(\xi, \eta_1, \eta_2)J^{-1}(\xi^*, \eta_1^*, \eta_2^*)'}{\sqrt{(\xi, \eta_1, \eta_2)J^{-1}(\xi, \eta_1, \eta_2)'(\xi^*, \eta_1^*, \eta_2^*)J^{-1}(\xi^*, \eta_1^*, \eta_2^*)'}}.$$

The condition for the orthogonality of the two spheres is therefore

$$(\xi, \eta_1, \eta_2)J^{-1}(\xi^*, \eta_1^*, \eta_2^*)' = 0. \tag{1}$$

Definition. The set of spheres orthogonal to a given sphere is called a *family of spheres.*

Thus, depending on whether the given sphere in the preceding definition is the real sphere, the point sphere or the imaginary sphere, we obtain a hyperbolic family, a parabolic family or an elliptic family.

3.6 Conformal Mappings

In this section we shall discuss some even more general results.
Suppose we have a transformation

$$y = y(x),$$

such that

$$dy\, dy' = \frac{1}{\lambda^2}\, dx\, dx', \qquad \lambda = \lambda(x), \tag{1}$$

and such that it transforms the domain D_x into the domain D_y. Then this transformation is called a *conformal mapping which takes D_x into D_y.* Concerning the Laplacian we have the following property:

Theorem 1. *If the transformation $y = y(x)$ satisfies* (1), *and*

$$Y(y) = X(x)\lambda^{n/2 - 1}, \tag{2}$$

then

$$\sum_{i=1}^{n} \frac{\partial^2 Y}{\partial y_i^2} = \lambda^{n/2 + 1} \sum_{i=1}^{n} \frac{\partial^2 X}{\partial x_i^2}. \tag{3}$$

PROOF. (a) From (1) we obtain

$$\sum_{i=1}^{n} \frac{\partial y_i}{\partial x_p} \frac{\partial y_i}{\partial x_q} = \frac{1}{\lambda^2} \delta_{pq}, \qquad \sum_{i=1}^{n} \frac{\partial x_i}{\partial y_p} \frac{\partial x_i}{\partial y_q} = \lambda^2 \delta_{pq}, \tag{4}$$

and moreover,

$$\frac{\partial x_i}{\partial y_p} = \lambda^2 \frac{\partial y_p}{\partial x_i}. \tag{5}$$

(b) Differentiating (4) gives

$$\sum_{i=1}^{n} \frac{\partial^2 y_i}{\partial x_p \partial x_s} \cdot \frac{\partial y_i}{\partial x_q} + \sum_{i=1}^{n} \frac{\partial y_i}{\partial x_p} \cdot \frac{\partial^2 y_i}{\partial x_q \partial x_s} = \delta_{pq} \frac{\partial \lambda^{-2}}{\partial x_s}. \tag{6}$$

Interchanging the summation indices s and q gives

$$\sum_{i=1}^{n} \frac{\partial^2 y_i}{\partial x_p \partial x_q} \cdot \frac{\partial y_i}{\partial x_s} + \sum_{i=1}^{n} \frac{\partial y_i}{\partial x_p} \cdot \frac{\partial^2 y_i}{\partial x_q \partial x_s} = \delta_{ps} \frac{\partial \lambda^{-2}}{\partial x_q}. \tag{7}$$

Then adding (6) and (7) we obtain

$$\frac{\partial}{\partial x_p} \left(\sum_{i=1}^{n} \frac{\partial y_i}{\partial x_s} \frac{\partial y_i}{\partial x_q} \right) + 2 \sum_{i=1}^{n} \frac{\partial y_i}{\partial x_p} \frac{\partial^2 y_i}{\partial x_q \partial x_s} = \delta_{pq} \frac{\partial \lambda^{-2}}{\partial x_s} + \delta_{ps} \frac{\partial \lambda^{-2}}{\partial x_q},$$

i.e.

$$2 \sum_{i=1}^{n} \frac{\partial y_i}{\partial x_p} \cdot \frac{\partial^2 y_i}{\partial x_q \partial x_s} = \delta_{pq} \frac{\partial \lambda^{-2}}{\partial x_s} + \delta_{ps} \frac{\partial \lambda^{-2}}{\partial x_q} - \delta_{sq} \frac{\partial \lambda^{-2}}{\partial x_p}. \tag{8}$$

Multiplying by $\partial x_p / \partial y_j$ and summing over p, we have

$$2 \frac{\partial^2 y_j}{\partial x_q \partial x_s} = \frac{\partial \lambda^{-2}}{\partial x_s} \frac{\partial x_q}{\partial y_j} + \frac{\partial \lambda^{-2}}{\partial x_q} \frac{\partial x_s}{\partial y_j} - \delta_{sq} \frac{\partial \lambda^{-2}}{\partial y_j}. \tag{9}$$

(c) Letting $q = s$ and summing over q gives

$$2 \sum_{q=1}^{n} \frac{\partial^2 y_j}{\partial x_q^2} = (2 - n) \frac{\partial \lambda^{-2}}{\partial y_j}. \tag{10}$$

(d) In turn, taking the partial derivative of (8) with respect to x_t we have

$$2 \sum_{i=1}^{n} \frac{\partial^2 y_i}{\partial x_p \partial x_t} \cdot \frac{\partial^2 y_i}{\partial x_q \partial x_s} + 2 \sum_{i=1}^{n} \frac{\partial y_i}{\partial x_p} \cdot \frac{\partial^3 y_j}{\partial x_q \partial x_s \partial x_t}$$
$$= \delta_{pq} \frac{\partial^2 \lambda^{-2}}{\partial x_s \partial x_t} + \delta_{ps} \frac{\partial^2 \lambda^{-2}}{\partial x_q \partial x_t} - \delta_{sq} \frac{\partial^2 \lambda^{-2}}{\partial x_p \partial x_t}. \tag{11}$$

Interchanging the summation indices p and t gives

$$2 \sum_{i=1}^{n} \frac{\partial^2 y_i}{\partial x_p \partial x_t} \cdot \frac{\partial^2 y_i}{\partial x_q \partial x_s} + 2 \sum_{i=1}^{n} \frac{\partial y_i}{\partial x_t} \frac{\partial^3 y_i}{\partial x_q \partial x_s \partial x_p}$$
$$= \delta_{tq} \frac{\partial^2 \lambda^{-2}}{\partial x_s \partial x_p} + \delta_{ts} \frac{\partial^2 \lambda^{-2}}{\partial x_p \partial x_q} - \delta_{sq} \frac{\partial^2 \lambda^{-2}}{\partial x_p \partial x_t}. \tag{12}$$

Adding the two expressions:

$$4 \sum_{i=1}^{n} \frac{\partial^2 y_i}{\partial x_p \partial x_i} \frac{\partial^2 y_i}{\partial x_q \partial x_s} + 2 \left(\sum_{i=1}^{n} \frac{\partial y_i}{\partial x_p} \frac{\partial^3 y_i}{\partial x_q \partial x_s \partial x_t} + \sum_{i=1}^{n} \frac{\partial y_i}{\partial x_t} \frac{\partial^3 y_i}{\partial x_q \partial x_s \partial x_p} \right)$$
$$= \delta_{pq} \frac{\partial^2 \lambda^{-2}}{\partial x_s \partial x_t} + \delta_{ps} \frac{\partial^2 \lambda^{-2}}{\partial x_q \partial x_t} + \delta_{tq} \frac{\partial^2 \lambda^{-2}}{\partial x_s \partial x_p} + \delta_{ts} \frac{\partial^2 \lambda^{-2}}{\partial x_p \partial x_q} - 2\delta_{sq} \frac{\partial^2 \lambda^{-2}}{\partial x_p \partial x_t}. \tag{13}$$

Utilizing

$$\frac{\partial^2}{\partial x_q \partial x_s}\left(\sum_{i=1}^{n} \frac{\partial y_i}{\partial x_p}\frac{\partial y_i}{\partial x_t}\right) = \frac{\partial^2 \lambda^{-2}}{\partial x_q \partial x_s}\delta_{pt} \tag{14}$$

and

$$\frac{\partial^2}{\partial x_q \partial x_s}\left(\sum_{i=1}^{n} \frac{\partial y_i}{\partial x_p}\frac{\partial y_i}{\partial x_t}\right) = \sum_{i=1}^{n}\left(\frac{\partial y_i}{\partial x_p}\frac{\partial^3 y_i}{\partial x_q \partial x_s \partial x_t} + \frac{\partial y_i}{\partial x_t}\cdot\frac{\partial^3 y_i}{\partial x_q \partial x_s \partial x_p}\right.$$

$$\left. + \frac{\partial^2 y_i}{\partial x_p \partial x_q}\cdot\frac{\partial^2 y_i}{\partial x_s \partial x_t} + \frac{\partial^2 y_i}{\partial x_p \partial x_s}\cdot\frac{\partial^2 y_i}{\partial x_q \partial x_t}\right), \tag{15}$$

then from (13), (14) and (15) we obtain

$$4\sum_{i=1}^{n}\frac{\partial^2 y_i}{\partial x_p \partial x_t}\frac{\partial^2 y_i}{\partial x_q \partial x_s} + 2\left(\frac{\partial^2 \lambda^{-2}}{\partial x_s \partial x_q}\delta_{pt} - \sum_{i=1}^{n}\frac{\partial^2 y_i}{\partial x_p \partial x_q}\cdot\frac{\partial^2 y_i}{\partial x_s \partial x_t}\right.$$

$$\left.- \sum_{i=1}^{n}\frac{\partial^2 y_i}{\partial x_p \partial x_s}\cdot\frac{\partial^2 y_i}{\partial x_q \partial x_t}\right) = \delta_{pq}\frac{\partial^2 \lambda^{-2}}{\partial x_s \partial x_t} + \delta_{ps}\frac{\partial^2 \lambda^{-2}}{\partial x_q \partial x_t} + \delta_{tq}\frac{\partial^2 \lambda^{-2}}{\partial x_s \partial x_p}$$

$$+ \delta_{ts}\frac{\partial^2 \lambda^{-2}}{\partial x_p \partial x_q} - 2\delta_{sq}\frac{\partial^2 \lambda^{-2}}{\partial x_p \partial x_t},$$

that is,

$$4\sum_{i=1}^{n}\frac{\partial^2 y_i}{\partial x_p \partial x_t}\cdot\frac{\partial^2 y_i}{\partial x_q \partial x_s} - 2\sum_{i=1}^{n}\frac{\partial^2 y_i}{\partial x_p \partial x_q}\frac{\partial^2 y_i}{\partial x_s \partial x_t} - 2\sum_{i=1}^{n}\frac{\partial^2 y_i}{\partial x_p \partial x_s}\frac{\partial^2 y_i}{\partial x_q \partial x_t}$$

$$= \delta_{pq}\frac{\partial^2 \lambda^{-2}}{\partial x_s \partial x_t} + \delta_{ps}\frac{\partial^2 \lambda^{-2}}{\partial x_q \partial x_t} + \delta_{tq}\frac{\partial^2 \lambda^{-2}}{\partial x_s \partial x_p} + \delta_{ts}\frac{\partial^2 \lambda^{-2}}{\partial x_p \partial x_q}$$

$$- 2\delta_{sq}\frac{\partial^2 \lambda^{-2}}{\partial x_p \partial x_t} - 2\delta_{pt}\frac{\partial^2 \lambda^{-2}}{\partial x_s \partial x_q}. \tag{16}$$

(e) In (16) choose $t = p$, $s = q$; then we get

$$4\sum_{i=1}^{n}\frac{\partial^2 y_i}{\partial x_p^2}\cdot\frac{\partial^2 y_i}{\partial x_q^2} - 4\sum_{i=1}^{n}\left(\frac{\partial^2 y_i}{\partial x_p \partial x_q}\right)^2 = 4\delta_{pq}\frac{\partial^2 \lambda^{-2}}{\partial x_p \partial x_q} - 2\frac{\partial^2 \lambda^{-2}}{\partial x_p^2} - 2\frac{\partial^2 \lambda^{-2}}{\partial x_q^2}.$$

Summing over p and q,

$$\sum_{i=1}^{n}\left[\left(\sum_{p}\frac{\partial^2 y_i}{\partial x_p^2}\right)^2 - \sum_{p,q}\left(\frac{\partial^2 y_i}{\partial x_p \partial x_q}\right)^2\right] = (1 - n)\sum_{p=1}^{n}\frac{\partial^2 \lambda^{-2}}{\partial x_p^2}. \tag{17}$$

(f) We know from (9) that

$$4\left(\frac{\partial^2 y_i}{\partial x_p \partial x_q}\right)^2 = \left(\frac{\partial \lambda^{-2}}{\partial x_p}\frac{\partial x_q}{\partial y_i} + \frac{\partial \lambda^{-2}}{\partial x_q}\frac{\partial x_p}{\partial y_i} - \delta_{pq}\frac{\partial \lambda^{-2}}{\partial y_i}\right)^2,$$

and therefore

$$
\begin{aligned}
4 \sum_i \sum_{p,q} \left(\frac{\partial^2 y_i}{\partial x_p \partial x_q} \right)^2 &= 4 \sum_{p,q} \sum_i \left(\frac{\partial^2 y_i}{\partial x_p \partial x_q} \right)^2 \\
&= \sum_{p,q} \sum_i \left[\left(\frac{\partial \lambda^{-2}}{\partial x_p} \right)^2 \left(\frac{\partial x_q}{\partial y_i} \right)^2 + \left(\frac{\partial \lambda^{-2}}{\partial x_q} \right)^2 \left(\frac{\partial x_p}{\partial y_i} \right)^2 \right. \\
&\quad + \delta_{pq} \left(\frac{\partial \lambda^{-2}}{\partial y_i} \right)^2 + 2 \frac{\partial \lambda^{-2}}{\partial x_p} \frac{\partial \lambda^{-2}}{\partial x_q} \frac{\partial x_q}{\partial y_i} \frac{\partial x_p}{\partial y_i} \\
&\quad \left. - 2 \frac{\partial \lambda^{-2}}{\partial x_p} \cdot \frac{\partial \lambda^{-2}}{\partial y_i} \delta_{pq} \frac{\partial x_q}{\partial y_i} - 2 \frac{\partial \lambda^{-2}}{\partial x_q} \frac{\partial \lambda^{-2}}{\partial y_i} \delta_{pq} \frac{\partial x_p}{\partial y_i} \right] \\
&= 2n\lambda^2 \sum_p \left(\frac{\partial \lambda^{-2}}{\partial x_p} \right)^2 + n \sum_i \left(\frac{\partial \lambda^{-2}}{\partial y_i} \right)^2 + 2\lambda^2 \sum_p \left(\frac{\partial \lambda^{-2}}{\partial x_p} \right)^2 \\
&\quad - 4 \sum_i \left(\frac{\partial \lambda^{-2}}{\partial y_i} \right)^2 = 2(n+1)\lambda^2 \sum_p \left(\frac{\partial \lambda^{-2}}{\partial x_p} \right)^2 \\
&\quad + (n-4) \sum_i \left(\frac{\partial \lambda^{-2}}{\partial y_i} \right)^2 = (3n-2)\lambda^2 \sum_p \left(\frac{\partial \lambda^{-2}}{\partial x_p} \right)^2, \quad (18)
\end{aligned}
$$

where we have made use of the following:

$$
\begin{aligned}
\sum_{i=1}^n \left(\frac{\partial \lambda^{-2}}{\partial y_i} \right)^2 &= \sum_{i=1}^n \left(\sum_j \frac{\partial \lambda^{-2}}{\partial x_j} \frac{\partial x_j}{\partial y_i} \right)^2 \\
&= \sum_{j,k} \frac{\partial \lambda^{-2}}{\partial x_j} \frac{\partial \lambda^{-2}}{\partial x_k} \sum_i \frac{\partial x_j}{\partial y_i} \frac{\partial x_k}{\partial y_i} \\
&= \lambda^2 \sum_{j=1}^n \left(\frac{\partial \lambda^{-2}}{\partial x_j} \right)^2. \quad (19)
\end{aligned}
$$

(g) Substituting (10) and (18) into (17) (and then using (19)) yields

$$
\sum_{j=1}^n \left[\left(1 - \frac{n}{2} \right)^2 \frac{\partial \lambda^{-2}}{\partial x_j} \right]^2 - \frac{3n-2}{4} \lambda^2 \sum_p \left(\frac{\partial \lambda^{-2}}{\partial x_p} \right)^2 = (1-n) \sum_p \frac{\partial^2 \lambda^{-2}}{\partial x_p^2},
$$

i.e.

$$
\left[\left(1 - \frac{n}{2} \right)^2 - \frac{3n-2}{4} \right] \lambda^2 \sum_p \left(\frac{\partial \lambda^{-2}}{\partial x_p} \right)^2 = (1-n) \sum_p \frac{\partial^2 \lambda^{-2}}{\partial x_p^2},
$$

$$
\frac{(1-n)(6-n)}{4} \lambda^2 \sum_p \left(\frac{\partial \lambda^{-2}}{\partial x_p} \right)^2 = (1-n) \sum_p \frac{\partial^2 \lambda^{-2}}{\partial x_p^2},
$$

or, simplifying,

$$(6 - n)\lambda^2 \sum_p \left(\frac{\partial \lambda^{-2}}{\partial x_p}\right)^2 = 4 \sum_p \frac{\partial^2 \lambda^{-2}}{\partial x_p^2},$$

so that, finally,

$$\frac{n}{2} \sum_p \left(\frac{\partial \lambda}{\partial x_p}\right)^2 = \lambda \sum_p \frac{\partial^2 \lambda}{\partial x_p^2}. \tag{20}$$

(h) From

$$\frac{\partial Y}{\partial y_i} = \sum_{j=1}^n \frac{\partial Y}{\partial x_j} \frac{\partial x_j}{\partial y_i}$$

and

$$\frac{\partial^2 Y}{\partial y_i^2} = \sum_{j,k=1}^n \frac{\partial^2 Y}{\partial x_j \partial x_k} \cdot \frac{\partial x_j}{\partial y_i} \cdot \frac{\partial x_k}{\partial y_i} + \sum_{j=1}^n \frac{\partial Y}{\partial x_j} \frac{\partial^2 x_j}{\partial y_i^2},$$

as well as (10), we know that

$$\sum_{i=1}^n \frac{\partial^2 Y}{\partial y_i^2} = \lambda^2 \sum_{i=1}^n \frac{\partial^2 Y}{\partial x_i^2} + \sum_{j=1}^n \frac{\partial Y}{\partial x_j} \sum_{i=1}^n \frac{\partial^2 x_j}{\partial y_i^2}$$

$$= \lambda^2 \sum_{i=1}^n \frac{\partial^2 Y}{\partial x_i^2} + (2 - n)\lambda \sum_{j=1}^n \frac{\partial Y}{\partial x_j} \frac{\partial \lambda}{\partial x_j}$$

$$= \lambda^n \sum_{i=1}^n \frac{\partial}{\partial x_i} \left(\lambda^{2-n} \frac{\partial Y}{\partial x_i}\right).$$

(i) Making use of the identity

$$\frac{\partial}{\partial x_i} \left(\Omega^2 \frac{\partial}{\partial x} (\Omega^{-1} \Phi)\right) = \Omega \frac{\partial^2 \Phi}{\partial x_i^2} - \Phi \frac{\partial^2 \Omega}{\partial x_i^2},$$

we obtain

$$\sum_{i=1}^n \frac{\partial^2 Y}{\partial y_i^2} = \lambda^n \sum_{i=1}^n \frac{\partial}{\partial x_i} \left(\lambda^{2-n} \frac{\partial(\lambda^{n/2-1} X)}{\partial x_i}\right)$$

$$= \sum_{i=1}^n \lambda^n \left(\lambda^{1-n/2} \frac{\partial^2 X}{\partial x_i^2} - X \frac{\partial^2 \lambda^{1-n/2}}{\partial x_i^2}\right)$$

$$= \lambda^{n/2+1} \sum_{i=1}^n \frac{\partial^2 X}{\partial x_i^2} - X\lambda^n \sum_{i=1}^n \frac{\partial^2 \lambda^{1-n/2}}{\partial x_i^2}$$

$$= \lambda^{n/2+1} \sum_{i=1}^n \frac{\partial^2 X}{\partial x_i^2} - X\left(1 - \frac{n}{2}\right)\lambda^{n/2-1}\left(\lambda \sum_i \frac{\partial^2 \lambda}{\partial x_i^2} - \frac{n}{2} \sum_i \left(\frac{\partial \lambda}{\partial x_i}\right)^2\right),$$

and from (20) we see that the last term equals 0. Therefore,

$$\sum_{i=1}^{n} \frac{\partial^2 Y}{\partial y_i^2} = \lambda^{n/2+1} \sum_{i=1}^{n} \frac{\partial^2 X}{\partial x_i^2}.$$

Remark. Note that in the preceding proof we have also proved along the way that

$$\lambda^{1-1/2n}, \qquad \lambda^{1-1/2n} y_i, \qquad i = 1, \dots, n$$

are harmonic functions.

Question 1[1]. Does a conformal mapping which carries the unit ball onto itself necessarily decrease non-Euclidean distance?

Question 2[1]. If a conformal mapping which carries the unit ball onto itself leaves the origin fixed, does it also carry each concentric ball into itself?

---

[1] *Translator's note*: Professor Lu Qikeng (K. H. Look) has kindly informed us that these questions have been settled affirmatively by two of his students, Yang Hongcang and Hong Yi, in their paper, *On conformal mappings of hyperballs* (Chinese), (to appear).

CHAPTER 4
The Lorentz Group

4.1 Changing the Basic Square Matrix

We have previously defined

$$J = \begin{pmatrix} I^{(n)} & 0 & 0 \\ 0 & 0 & -\frac{1}{2} \\ 0 & -\frac{1}{2} & 0 \end{pmatrix}, \tag{1}$$

and we have studied the group of all M satisfying

$$MJM' = J. \tag{2}$$

Since

$$\begin{pmatrix} 1 & -1 \\ 1 & 1 \end{pmatrix} \begin{pmatrix} 0 & -\frac{1}{2} \\ -\frac{1}{2} & 0 \end{pmatrix} \begin{pmatrix} 1 & -1 \\ 1 & 1 \end{pmatrix}' = \begin{pmatrix} 0 & 0 \\ 0 & -1 \end{pmatrix},$$

if we let

$$P = \begin{pmatrix} I^{(n)} & 0 \\ 0 & \begin{pmatrix} 1 & -1 \\ 1 & 1 \end{pmatrix} \end{pmatrix} M \begin{pmatrix} I^{(n)} & 0 \\ 0 & \begin{pmatrix} 1 & -1 \\ 1 & 1 \end{pmatrix} \end{pmatrix}^{-1}, \tag{3}$$

then

$$P \begin{pmatrix} I^{(n+1)} & 0 \\ 0 & -1 \end{pmatrix} P' = \begin{pmatrix} I^{(n+1)} & 0 \\ 0 & -1 \end{pmatrix}. \tag{4}$$

(3) may be written more concretely:

$$P = \begin{pmatrix} M_1 & \frac{1}{2}(u_1' - u_2') & \frac{1}{2}(u_1' + u_2') \\ v_1 - v_2 & \frac{1}{2}(a - b - c + d) & \frac{1}{2}(a + b - c - d) \\ v_1 + v_2 & \frac{1}{2}(a - b + c - d) & \frac{1}{2}(a + b + c + d) \end{pmatrix},$$

in which

$$M = \begin{pmatrix} M_1 & u_1' & u_2' \\ v_1 & a & b \\ v_2 & c & d \end{pmatrix}.$$

For quite a long while, we shall be studying the group of all P satisfying (4). This group is denoted by $L(n + 1, 1)$ and it is called the Lorentz group of type $(n + 1, 1)$.

From (4) we have

$$(\det P)^2 = 1,$$

that is, $\det P = \pm 1$. Those P whose determinant is equal to 1 form a group, to be denoted by $L_{(n+1, 1)}^+$; the subset of those P whose determinant is equal to -1 will be denoted by $L_{(n+1, 1)}^-$. In particular, if we denote by $[1, \dots, 1, -1]$ the diagonal matrix with $\{1, \dots, 1, -1\}$ on the diagonal, then it is a square matrix in $L_{(n+1, 1)}$ whose determinant is -1, and $L_{(n+1, 1)}^-$ is obtained from $L_{(n+1, 1)}^+$ by multiplying the square matrices of $L_{(n+1, 1)}^+$ with $[1, \dots, 1, -1]$, i.e.,

$$L_{(n+1, 1)}^- = [1, \dots, 1, -1]L_{(n+1, 1)}^+.$$

Thus the group $L_{(n+1, 1)}$ is generated by $L_{(n+1, 1)}^+$ and an arbitrary matrix in $L_{(n+1, 1)}$ with determinant equal to -1. Moreover,

$$\begin{aligned} L_{(n+1, 1)} &= L_{(n+1, 1)}^+ \cup L_{(n+1, 1)}^- \\ &= L_{(n+1, 1)}^+ \cup [1, \dots, 1, -1]L_{(n+1, 1)}^+. \end{aligned} \tag{5}$$

Now if such a P is expressed as

$$(a_{ij})_{1 \leqslant i, j \leqslant n+2},$$

then the matrices satisfying

$$a_{n+2, n+2} > 0 \tag{6}$$

form a group, to be denoted by $L_{+(n+1, 1)}$. The fact that the matrices which satisfy (6) is indeed a group can be proved as follows. Let

$$B = (b_{ij})_{1 \leqslant i, j \leqslant n+2}, \qquad b_{n+2, n+2} > 0$$

and

$$C = AB,$$

then

$$c_{n+2, n+2} = \sum_{i=1}^{n+1} a_{n+2, i}b_{i, n+2} + a_{n+2, n+2}b_{n+2, n+2}.$$

From relation (4), we already know that

$$\sum_{i=1}^{n+1} a_{n+2, i}^2 - a_{n+2, n+2}^2 = -1, \qquad \sum_{i=1}^{n+1} a_{n+2, i}^2 < a_{n+2, n+2}^2,$$

$$\sum_{i=1}^{n+1} b_{i, n+2}^2 - b_{n+2, n+2}^2 = -1, \qquad \sum_{i=1}^{n+1} b_{i, n+2}^2 < b_{n+2, n+2}^2,$$

and from the Schwarz inequality,

$$\left| \sum_{i=1}^{n+1} a_{n+2,i} b_{i,n+2} \right| \le \sqrt{\sum_{i=1}^{n+1} a_{n+2,i}^2 \sum_{i=1}^{n+1} b_{i,n+2}^2} < a_{n+2,n+2} b_{n+2,n+2},$$

which is what we wished to prove. The elements satisfying

$$a_{n+2,n+2} < 0 \tag{7}$$

form a set; this set is denoted by $L_{-(n+1,1)}$. Clearly we also have that $[1, \ldots, 1, -1] \in L_{-(n+1,1)}$, and that

$$L_{-(n+1,1)} = [1, \ldots, 1, -1] L_{+(n+1,1)};$$

moreover,

$$L_{(n+1,1)} = L_{+(n+1,1)} \cup L_{-(n+1,1)}$$
$$= L_{+(n+1,1)} \cup [1, \ldots, 1, -1] L_{+(n+1,1)}.$$

Those elements belonging to both $L_{(n+1,1)}^+$ and $L_{+(n+1,1)}$ form a group, to be denoted by $L_{+(n+1,1)}^+$, i.e.

$$L_{+(n+1,1)}^+ = L_{(n+1,1)}^+ \cap L_{+(n+1,1)}.$$

It is not difficult to prove that

$$L_{(n+1,1)}^+ = L_{+(n+1,1)}^+ \cup L_{-(n+1,1)}^+$$
$$= L_{+(n+1,1)}^+ \cup [-1, 1, \ldots, 1, -1] L_{+(n+1,1)}^+,$$

and that

$$L_{(n+1,1)} = L_{+(n+1,1)}^+ \cup L_{-(n+1,1)}^+ \cup L_{+(n+1,1)}^- \cup L_{-(n+1,1)}^-,$$

wherein

$$L_{-(n+1,1)}^+ = [-1, 1, \ldots, 1, -1] L_{+(n+1,1)}^+,$$
$$L_{+(n+1,1)}^- = [1, \ldots, 1, -1, 1] L_{+(n+1,1)}^+,$$
$$L_{-(n+1,1)}^- = [1, \ldots, 1, +1, -1] L_{+(n+1,1)}^+.$$

Note: In terms of the original notation, condition (6) is the statement that

$$a + b + c + d > 0.$$

4.2 Generators

Definition. Fix i, j ($i \ne j$, $1 \le i$, $j \le n + 1$). Then the $(n + 2) \times (n + 2)$ matrix corresponding to the transformation $(x_1, \ldots, x_{n+2}) \to (y_1, \ldots, y_{n+2})$ such that $y_i = x_j$, $y_j = x_i$ and $y_k = x_k$ for all $k \ne i, j$ is called a *permutation transformation*. It will be denoted by P_{ij} ($1 \le i, j \le n + 1$).

Thus $P_{ij} P$ is the matrix obtained from P by interchanging the ith and jth rows of P, and PP_{ij} is that interchanging the ith and jth columns of P. P_{ij} belongs to $L_{+(n+1,1)}^-$.

In the context of the preceding definition, suppose instead $y_i = x_j$, $y_j = -x_i$ and $y_k = x_k$ for all $k \neq i, j$; the resulting permutation transformation will be denoted by Q_{ij}. Clearly $Q_{ij} \in L^+_{+(n+1, 1)}$.

The transformation

$$R_{12}(\theta) \equiv \begin{pmatrix} \cos\theta & \sin\theta & 0 & \cdots & 0 \\ -\sin\theta & \cos\theta & 0 & \cdots & 0 \\ 0 & \cdots & 0 & 1 & \cdots & 0 \\ \vdots & & & & \ddots & \\ 0 & \cdots & 0 & 0 & \cdots & 1 \end{pmatrix}$$

is called a *rotation*, and it belongs to $L^+_{+(n+1, 1)}$. Now let

$$R_{ij}(\theta) = Q_{i1} Q_{i2} R_{12} Q_{j2}^{-1} Q_{j1}^{-1}, \qquad i, j = 1, 2, \ldots, n+1.$$

Then the transformation representing it is

$$y_i = \cos\theta x_i + \sin\theta x_j, \qquad y_j = -\sin\theta x_i + \cos\theta x_j.$$

The other x_k ($k \neq i, j$) are unchanged. These transformations are all referred to as *rotations*. We also define

$$H_1(\psi) \equiv \begin{pmatrix} \cosh\psi & 0 & \cdots & 0 & \sinh\psi \\ 0 & 1 & \cdots & 0 & 0 \\ \vdots & \vdots & \ddots & \vdots & \vdots \\ 0 & 0 & \cdots & 1 & 0 \\ \sinh\psi & 0 & \cdots & 0 & \cosh\psi \end{pmatrix}$$

to be a *hyperbolic rotation*. And

$$H_i(\psi) = Q_{i1} H_1 Q_{i1}^{-1}$$

is also called a *hyperbolic rotation*.

Theorem 1. $L^+_{+(n+1, 1)}$ *is generated by permutations, rotations and hyperbolic rotations.*

PROOF. (a) We begin with the following very simple property: let

$$\begin{pmatrix} a & b \\ c & d \end{pmatrix} \begin{pmatrix} \cosh\psi & \sinh\psi \\ \sinh\psi & \cosh\psi \end{pmatrix} = \begin{pmatrix} * & * \\ c' & * \end{pmatrix},$$

then by a direct calculation,

$$c' = c \cosh\psi + d \sinh\psi.$$

If $d \neq 0$ and $|c| \leqslant |d|$, then from

$$c \cosh\psi + d \sinh\psi = 0,$$

that is, from

$$\tanh\psi = -c/d,$$

it is possible to determine ψ. Thus there exists a ψ such that $c' = 0$.

(b) Let A be any matrix in $L^+_{+(n+1,\,1)}$, and write

$$A = \begin{pmatrix} A_1^{(n+1)} & \beta \\ \alpha & a_{n+2,\,n+2} \end{pmatrix}.$$

From $A[1,\ldots,1,-1]A' = [1,\ldots,1,-1]$ we know that

$$\alpha\alpha' + 1 = a^2_{n+2,\,n+2}.$$

Therefore

$$a_{n+2,\,n+2} \neq 0, \qquad |a_{n+2,\,j}| \leqslant |a_{n+2,\,n+2}|,$$
$$j = 1, 2, \ldots, n+1.$$

So there exists a hyperbolic rotation, $H_1(\psi_1)$, such that the $(n+2,1)$th entry of the matrix

$$AH_1(\psi_1)$$

is equal to 0, and such that $AH_1(\psi_1)$ still belongs to $L^+_{+(n+1,\,1)}$. We may continue in a similar manner, e.g. by multiplying by a suitable $H_2(\psi_2)$ we have the $(n+2,2)$th entry equal to 0. In other words, by multiplying with a hyperbolic rotation, A may be written in the form

$$\begin{pmatrix} \tilde{A}_1^{(n+1)} & \tilde{\beta} \\ 0 & \tilde{a}_{n+2,\,n+2} \end{pmatrix}.$$

Since this satisfies (1.4), we have $\tilde{\beta} = 0$, i.e.

$$\tilde{a}_{1,\,n+2} = \cdots = \tilde{a}_{n+1,\,n+2} = 0.$$

So A becomes

$$B = \begin{pmatrix} B_1^{(n+1)} & 0 \\ 0 & a \end{pmatrix}, \qquad B_1 B_1' = I^{(n+1)}, \qquad a^2 = 1.$$

Since B belongs to $L^+_{+(n+1,\,1)}$, we have $a = 1$, $\det B_1 = 1$. Consequently, B is an orthogonal matrix whose determinant is equal to 1, and we know that it is a product of rotations. (Those readers not familiar with this result may derive this result by using the steps discussed above.) □

To be more precise, $L^+_{+(n+1,\,1)}$ may be generated by permutations Q_{ij}, $R_{12}(\theta)$ and $H_1(\psi)$. Altogether there are $\frac{1}{2}n(n+1)$ permutations; note that they may be generated by

$$Q_{12} \text{ and } \begin{pmatrix} 0 & 1 & 0 & \cdots & 0 \\ 0 & 0 & 1 & 0 & \cdots & 0 \\ \multicolumn{6}{c}{\cdots\cdots\cdots\cdots\cdots\cdots} \\ 0 & 0 & 0 & 0 & \cdots & 1 \\ (-1)^n & 0 & 0 & 0 & \cdots & 0 \end{pmatrix}.$$

However, since $Q_{12} = R_{12}(\pi/2)$, we can see that altogether three types of elements will suffice to generate $L^+_{+(n+1,\,1)}$.

Question: Can the number of generators be further decreased?

Theorem 2. $L_{(n+1, 1)}$ *may be generated by* Q_{ij}, $R_{12}(\theta)$, $H_1(\psi)$, $[1, \ldots, 1, -1]$ *and* $[1, \ldots, 1, -1, -1]$.

Returning to the original notation, we have corresponding to R_{12}:

$$y_1 = \cos \theta x_1 + \sin \theta x_2,$$
$$y_2 = -\sin \theta x_1 + \cos \theta x_2,$$
$$y_i = x_i, \qquad i = 3, 4, \ldots, n+2$$

and corresponding to

$$\begin{pmatrix} 0 & 1 & 0 & \cdots & 0 \\ 0 & 0 & 1 & \cdots & 0 \\ \vdots & \vdots & \vdots & \ddots & \vdots \\ 0 & 0 & 0 & \cdots & 1 \\ (-1)^n & 0 & 0 & \cdots & 0 \end{pmatrix},$$

we have

$$u_1 = 2e_n, \qquad u_2 = -2e_n, \qquad v_1 = \tfrac{1}{2}(-1)^n e_1,$$
$$v_2 = -\tfrac{1}{2}(-1)^n e_1, \qquad a = b = c = d = \tfrac{1}{2},$$

i.e.

$$y_1 = (-1)^n \frac{xx' - 1}{-4x_n + xx' + 1},$$

$$y_i = \frac{2x_{i-1}}{-4x_n + xx' + 1}, \qquad i = 2, 3, \ldots, n.$$

Furthermore, corresponding to $H_1(\psi)$, we have

$$y_1 = \frac{(\cosh \psi)x_1}{(\sinh \psi)x_1 + \tfrac{1}{2}(1 - \cosh \psi)xx' + \tfrac{1}{2}(1 + \cosh \psi)},$$

$$y_i = \frac{x_i}{(\sinh \psi)x_1 + \tfrac{1}{2}(1 - \cosh \psi)xx' + \tfrac{1}{2}(1 + \cosh \psi)},$$

$$y_n = \frac{x_n}{(\sinh \psi)x_1 + \tfrac{1}{2}(1 - \cosh \psi)xx' + \tfrac{1}{2}(1 + \cosh \psi)}.$$

4.3 Orthogonal Similarity

For the purpose of introducing and solving the *Lorentz similarity problem*, we shall review the concept of orthogonal similarity, and the method of dealing with it.

Consider now the $m \times m$ orthogonal matrices, that is to say, the real square matrices T which satisfy

$$TT' = I. \tag{1}$$

Let A, B be two orthogonal matrices. If there exists an orthogonal square matrix T such that

$$TAT^{-1} = B, \tag{2}$$

then A, B are called *orthogonally similar*.

The geometric significance of orthogonal similarity is the following: Given an orthogonal transformation whose matrix relative to an orthonormal basis is A. If we transform this basis to a new orthonormal basis by matrix T, then with respect to the new orthonormal basis, the matrix of the orthogonal transformation is TAT^{-1}.

The most commonly known example is the case when $m = 3$. For any orthogonal transformation whose determinant is equal to 1, an orthonormal basis may be chosen such that the orthogonal transformation becomes a rotation around the z-axis, that is to say, there exists a T such that

$$TAT^{-1} = \begin{pmatrix} \cos\theta & \sin\theta & 0 \\ -\sin\theta & \cos\theta & 0 \\ 0 & 0 & 1 \end{pmatrix}.$$

We will now prove that any orthogonal matrix is orthogonally similar to

$$\begin{pmatrix} \cos\theta_1 & \sin\theta_1 \\ -\sin\theta_1 & \cos\theta_1 \end{pmatrix} \dotplus \begin{pmatrix} \cos\theta_2 & \sin\theta_2 \\ -\sin\theta_2 & \cos\theta_2 \end{pmatrix} \dotplus \cdots \dotplus \begin{pmatrix} \cos\theta_v & \sin\theta_v \\ -\sin\theta_v & \cos\theta_v \end{pmatrix}$$

$$\dotplus 1 \dotplus \cdots \dotplus 1 \dotplus (-1) \dotplus \cdots \dotplus (-1), \tag{3}$$

where $0 < \theta_1 \leqslant \theta_2 \leqslant \cdots \leqslant \theta_v < 2\pi$, and the rotation $\dotplus$ means direct sums, that is the summands are the square blocks along the diagonal of a matrix, whose entries elsewhere are equal to 0.

This is a well known result, but we nevertheless write down its proof as a model for subsequent discussions.

(a) We must first prove: given r mutually orthogonal unit vectors

$$v_1, \ldots, v_r \qquad (r < m),$$

i.e.

$$v_i v'_j = \delta_{ij},$$

we may add an additional v_{r+1}, i.e.

$$v_{r+1} v'_r = 0, \qquad v_{r+1} v'_{r+1} = 1.$$

This proof is extremely simple. Choose a linear vector u which is independent of $v_1, \ldots, v_r$. Let

$$u_{r+1} = u - c_1 v_1 - \cdots - c_r v_r, \qquad c_\gamma = u v'_\gamma,$$

then we obviously have

$$u_{r+1}v'_y = uv'_y - c_y = 0.$$

Since u_{r+1} is not the zero vector,

$$u_{r+1}u'_{r+1} = \xi > 0.$$

Then $v_{r+1} = 1/\xi^{1/2}u_{r+1}$ is the desired vector.   ☐

(b) The absolute value of any eigenvalue of an orthogonal matrix is equal to 1.

Let λ be an eigenvalue of A, and let its corresponding eigenvector be z, i.e.

$$zA = \lambda z.$$

Since A is a real matrix,

$$\bar{z}A = \bar{\lambda}\bar{z}.$$

Therefore,

$$z\bar{z}' = zAA'\bar{z}' = |\lambda|^2 z\bar{z}',$$

and $z\bar{z}' \neq 0$. Thus $|\lambda|^2 = 1$.

(c) If A has a complex eigenvalue $e^{i\theta}$ ($\neq \pm 1$), and

$$zA = e^{i\theta}z$$

divides z into its real and imaginary parts, namely $z = x + iy$, then

$$xA = (\cos \theta)x - (\sin \theta)y, \qquad yA = (\sin \theta)x + (\cos \theta)y,$$

that is,

$$\begin{pmatrix} x \\ y \end{pmatrix} A = \begin{pmatrix} \cos \theta & -\sin \theta \\ \sin \theta & \cos \theta \end{pmatrix} \begin{pmatrix} x \\ y \end{pmatrix}.$$

The 2×2 square matrix

$$\begin{pmatrix} x \\ y \end{pmatrix} \begin{pmatrix} x \\ y \end{pmatrix}' = \begin{pmatrix} xx' & xy' \\ yx' & yy' \end{pmatrix} = \begin{pmatrix} s & t \\ t & u \end{pmatrix}$$

is positive definite; moreover

$$\begin{pmatrix} x \\ y \end{pmatrix} \begin{pmatrix} x \\ y \end{pmatrix}' = \begin{pmatrix} x \\ y \end{pmatrix} AA' \begin{pmatrix} x \\ y \end{pmatrix}'$$

$$= \begin{pmatrix} \cos \theta & -\sin \theta \\ \sin \theta & \cos \theta \end{pmatrix} \begin{pmatrix} x \\ y \end{pmatrix} \begin{pmatrix} x \\ y \end{pmatrix}' \begin{pmatrix} \cos \theta & -\sin \theta \\ \sin \theta & \cos \theta \end{pmatrix}',$$

i.e.

$$\begin{pmatrix} s & t \\ t & u \end{pmatrix} = \begin{pmatrix} \cos \theta & \sin \theta \\ -\sin \theta & \cos \theta \end{pmatrix} \begin{pmatrix} s & t \\ t & u \end{pmatrix} \begin{pmatrix} \cos \theta & -\sin \theta \\ \sin \theta & \cos \theta \end{pmatrix}.$$

This then gives $t = 0$, $s = u$. Let

$$v_1 = \frac{1}{\sqrt{s}} x, \qquad v_2 = \frac{1}{\sqrt{s}} y,$$

then

$$\begin{pmatrix} v_1 \\ v_2 \end{pmatrix} A = \begin{pmatrix} \cos \theta & -\sin \theta \\ \sin \theta & \cos \theta \end{pmatrix} \begin{pmatrix} v_1 \\ v_2 \end{pmatrix},$$

and v_1, v_2 are two mutually orthogonal unit vectors.

From (a) we can construct a square matrix T which has v_1, v_2 as its first two rows. Then

$$TAT^{-1} = TAT' = \begin{pmatrix} \cos \theta & -\sin \theta & a_{13} & \cdots & a_{1m} \\ \sin \theta & \cos \theta & a_{23} & \cdots & a_{2m} \\ a_{31} & a_{32} & a_{33} & \cdots & a_{3m} \\ \cdots\cdots\cdots\cdots\cdots\cdots\cdots\cdots\cdots \\ a_{m1} & a_{m2} & a_{m3} & \cdots & a_{mm} \end{pmatrix}$$

(where we have used $\begin{pmatrix} v_1 \\ v_2 \end{pmatrix} A \begin{pmatrix} v_1 \\ v_2 \end{pmatrix}' = \begin{pmatrix} \cos \theta & -\sin \theta \\ \sin \theta & \cos \theta \end{pmatrix}$), and from the orthogonality of TAT',

$$a_{13} = \cdots = a_{1m} = a_{23} = \cdots = a_{2m} = 0,$$
$$a_{31} = \cdots = a_{m1} = a_{32} = \cdots = a_{m2} = 0.$$

By induction we obtain the desired result.

(d) If A has an eigenvalue equal to 1, let x be the corresponding eigenvector, i.e.

$$xA = x;$$

since we may assume that $xx' = 1$, construct an orthogonal matrix T which has x as its first row. Then

$$TAT^{-1} = \begin{pmatrix} 1 & 0 & \cdots & 0 \\ 0 & a_{22} & \cdots & a_{2m} \\ \cdots\cdots\cdots\cdots\cdots\cdots \\ 0 & a_{m2} & \cdots & a_{mm} \end{pmatrix},$$

and we may apply induction again.

The same method applies for the case when A has an eigenvalue equal to -1. $\qquad \square$

Remark. The necessary and sufficient condition for two orthogonal matrices to be orthogonally similar is that they have the same characteristic polynomial.

4.4 On Indefinite Quadratic Forms

We first prove several simple results about the quadratic form

$$x_1^2 + \cdots + x_{m-1}^2 - x_m^2 = xFx', \qquad F = [1, \ldots, 1, -1].$$

(a) It is impossible to have two linearly independent vectors y, z such that for all real numbers λ, μ,

$$(\lambda y + \mu z)F(\lambda y + \mu z)' = 0.$$

PROOF: This is equivalent to

$$\lambda^2 yFy' + 2\lambda\mu yFz' + \mu^2 zFz' = 0,$$

which is equivalent to

$$yFy' = 0, \qquad yFz' = 0, \qquad zFz' = 0,$$

which is equivalent to

$$y_1^2 + \cdots + y_{m-1}^2 = y_m^2,$$
$$z_1^2 + \cdots + z_{m-1}^2 = z_m^2,$$
$$y_1 z_1 + \cdots + y_{m-1} z_{m-1} = y_m z_m.$$

From this we have

$$(y_1^2 + \cdots + y_{m-1}^2)(z_1^2 + \cdots + z_{m-1}^2) - (y_1 z_1 + \cdots + y_{m-1} z_{m-1})^2 = 0,$$

i.e.

$$\sum_{i<j} (y_i z_j - y_j z_i)^2 = 0,$$

but this contradicts the linear independence of y, z. □

(b) Let P be an $l \times m$ matrix ($l < m$), then the index of

$$PFP'$$

has at most one negative sign.

PROOF. PFP' is an $l \times l$ square matrix. Now suppose the index has two negative signs, i.e. suppose there exists an $l \times l$ matrix Q such that

$$QPFP'Q' = \begin{pmatrix} -1 & & & \\ & -1 & & \\ & & \ddots & \end{pmatrix}.$$

Let the first and second rows of QP be y and z, respectively, then

$$yFy' = -1, \qquad zFz' = -1, \qquad yFz' = 0,$$

which is equivalent to

$$\sum_{i=1}^{m-1} y_i^2 = y_m^2 - 1, \qquad \sum_{i=1}^{m-1} z_i^2 = z_m^2 - 1, \qquad \sum_{i=1}^{m-1} y_i z_i = z_m y_m,$$

which is equivalent to

$$\left(\sum_{i=1}^{m-1} y_i^2 + 1\right)\left(\sum_{i=1}^{m-1} z_i^2 + 1\right) = \left(\sum_{i=1}^{m-1} y_i z_i\right)^2,$$

and this, by the same reasoning as in (a), is impossible. □

(c) The method of proof in the preceding sections actually gives the following result.

If the rank of P is equal to l, then the index of

$$PFP'$$

has only one of the following three possibilities:

(i) positive definite, i.e. $l + 1$'s;
(ii) semidefinite, i.e. $(l - 1) + 1$'s, one 0;
(iii) indefinite, i.e. $(l - 1) + 1$'s, one -1.

4.5 Lorentz Similarity

In the succeeding sections, let

$$F = \begin{pmatrix} I^{(n+1)} & 0 \\ 0 & -1 \end{pmatrix} \tag{1}$$

then a square matrix T satisfying

$$TFT' = F,$$

will be called simply a *Lorentz matrix*.

Let A, B be two Lorentz matrices; if there exists a Lorentz matrix T such that

$$TAT^{-1} = B,$$

then A, B are said to be *Lorentz similar*.

We shall prove the following: any Lorentz matrix is Lorentz similar to the direct sum of the following six types of square matrices:

$$\begin{pmatrix} \cos\theta, & \sin\theta \\ -\sin\theta, & \cos\theta \end{pmatrix}, +\mathbf{I}, -\mathbf{I}, \quad \begin{pmatrix} \cosh\psi, & \sinh\psi \\ \sinh\psi, & \cosh\psi \end{pmatrix},$$

$$\begin{pmatrix} 0 & 3 & 2\sqrt{2} \\ -1 & 0 & 0 \\ 0 & 2\sqrt{2} & 3 \end{pmatrix}, \quad \begin{pmatrix} 0 & -3 & -2\sqrt{2} \\ 1 & 0 & 0 \\ 0 & -2\sqrt{2} & -3 \end{pmatrix}.$$

Moreover, at most one of the last three types may appear, and then only once.

(a) Before proving this result, let us first prove the following:
If $v_1, \ldots, v_r$ satisfy

$$v_i F v_j' = \delta_{ij}, \qquad i, j = 1, \ldots, r,$$

then it is possible to add a v_{r+1} such that

$$v_{r+1} F v_j' = 0, \qquad j = 1, \ldots, r,$$

and

$$v_{r+1} F v_{r+1}' = -1.$$

PROOF. Let

$$u_{r+1} = e_n - \lambda_1 v_1 - \cdots - \lambda_r v_r, \qquad \lambda_\nu = e_n F v_\nu',$$

then

$$u_{r+1} F v_\nu' = e_n F v_\nu' - \lambda_\nu = 0.$$

But consider also

$$u_{r+1} F u_{r+1}' = e_n F e_n' - 2 \sum_{\nu=1}^{r} \lambda_\nu e_n F v_\nu' + \sum_{i,j} \lambda_i \lambda_j v_i F v_j'$$

$$= -1 - 2 \sum_{\nu=1}^{r} \lambda_\nu^2 + \sum_{\nu=1}^{r} \lambda_\nu^2$$

$$= -1 - \sum_{\nu=1}^{r} \lambda_\nu^2 < 0,$$

Then

$$v_{r+1} = u_{r+1} / \sqrt{1 + \lambda_1^2 + \cdots + \lambda_r^2}$$

furnishes the desired vector.   □

If $r + 1 < n$, then it is possible furthermore to add a vector v_{r+2} such that

$$v_{r+2} F v_\gamma' = 0, \qquad \gamma = 1, 2, \ldots, r + 1$$

and

$$v_{r+2} F v_{r+2}' = 1.$$

Indeed, choose a vector u which is linearly independent of $v_1, \ldots, v_{r+1}$, and set

$$u_{r+2} = u - \sum_{\nu=1}^{r+1} \lambda_\nu v_\nu, \qquad \lambda_\nu = u F v_\nu', \qquad \lambda_{r+1} = -u F v_{r+1}'.$$
$$(\nu = 1, \ldots, r)$$

Then

$$u_{r+2} F v_\gamma' = 0, \qquad \gamma = 1, \ldots, r + 1.$$

If

$$u_{r+2} F u_{r+2}' \leqslant 0,$$

then

$$\begin{pmatrix} u_1 \\ u_2 \\ \vdots \\ u_{r+2} \end{pmatrix} F \begin{pmatrix} u_1 \\ u_2 \\ \vdots \\ u_{r+2} \end{pmatrix}' = [1, \ldots 1, -1, -1].$$

But from §4 we know that this is impossible. Thus, we may set

$$v_{r+2} = u_{r+2}/\sqrt{u_{r+2}Fu'_{r+2}},$$

which gives the desired vector.   □

(b) Now we will prove that A cannot have a complex eigenvalue whose absolute value does not equal 1.

PROOF. If $\rho e^{i\theta}$ ($\rho \neq 1$, $\theta \neq 0$, π) is an eigenvalue, and z is the corresponding eigenvector, then

$$zA = \rho e^{i\theta} z.$$

Since

$$zFz' = zAFA'z' = \rho^2 zFz',$$

and $\rho \neq 1$, therefore

$$zFz' = 0.$$

Writing z as $x + iy$, we then obtain

$$xFx' + yFy' = 0, \qquad xFy' = 0. \tag{1}$$

Again, since

$$A = FA'^{-1}F,$$

$(1/\rho)e^{-i\theta}$ is also an eigenvalue, and its eigenvector is $w = u + iv$, i.e.

$$wA = (1/\rho e^{i\theta})w.$$

In a similar manner we have

$$uFu' + vFv' = 0, \qquad uFv' = 0. \tag{2}$$

Again, from

$$zFw' = zAFA'w' = e^{2i\theta}zFw',$$

we therefore have

$$zFw' = 0,$$

i.e.

$$xFu' + yFv' = 0, \qquad xFv' - yFu' = 0. \tag{3}$$

Thus the end result is that we obtain

$$\begin{pmatrix} x \\ y \\ u \\ v \end{pmatrix} F \begin{pmatrix} x \\ y \\ u \\ v \end{pmatrix}' = \begin{pmatrix} a & 0 & c & d \\ 0 & -a & d & -c \\ c & d & b & 0 \\ d & -c & 0 & -b \end{pmatrix}.$$

If $a \neq 0$, then

$$\left(-\begin{pmatrix} I & 0 \\ -\begin{pmatrix} c & -d \\ d & c \end{pmatrix}\begin{pmatrix} a & 0 \\ 0 & -a \end{pmatrix}^{-1} & I \end{pmatrix} \right) \begin{pmatrix} a & 0 & c & d \\ 0 & -a & d & -c \\ c & d & b & 0 \\ d & -c & 0 & -b \end{pmatrix}$$

$$\times \left(-\begin{pmatrix} I & 0 \\ -\begin{pmatrix} c & -d \\ d & c \end{pmatrix}\begin{pmatrix} a & 0 \\ 0 & -a \end{pmatrix}^{-1} & I \end{pmatrix} \right)' = \begin{pmatrix} a & 0 & 0 & 0 \\ 0 & -a & 0 & 0 \\ 0 & 0 & p & * \\ 0 & 0 & * & -p \end{pmatrix},$$

where $p = b - \dfrac{c^2 - d^2}{a}$,

and this is a symmetric matrix having two negative signs. When $a = 0$, it is also quite clear that it is a square matrix which neither has only one negative sign nor is singular. From the result of §4 we know that this is impossible. ☐

(c) If A has a real eigenvalue $\lambda \neq \pm 1$, then, since $A'^{-1} = FAF$, we therefore know that $1/\lambda$ is also an eigenvalue. Let x and y be the eigenvectors corresponding to λ and $1/\lambda$, respectively, i.e.

$$xA = \lambda x, \qquad yA = (1/\lambda)y.$$

From

$$xFx' = xAFA'x' = \lambda^2 xFx',$$

we know that

$$xFx' = 0,$$

and similarly, that

$$yFy' = 0.$$

Letting $xFy' = a$, then

$$\begin{pmatrix} x \\ y \end{pmatrix} F \begin{pmatrix} x \\ y \end{pmatrix}' = \begin{pmatrix} 0 & a \\ a & 0 \end{pmatrix}.$$

If $a = 0$, then from §4 we know that this is impossible. Choosing x/a to be the new x then yields

$$xFx' = yFy' = 0, \qquad xFy' = 1.$$

Since

$$\begin{pmatrix} \dfrac{1}{\sqrt{2}} & \dfrac{1}{\sqrt{2}} \\ -\dfrac{1}{\sqrt{2}} & \dfrac{1}{\sqrt{2}} \end{pmatrix} \begin{pmatrix} 0 & 1 \\ 1 & 0 \end{pmatrix} \begin{pmatrix} \dfrac{1}{\sqrt{2}} & \dfrac{1}{\sqrt{2}} \\ -\dfrac{1}{\sqrt{2}} & \dfrac{1}{\sqrt{2}} \end{pmatrix}' = \begin{pmatrix} 1 & 0 \\ 0 & -1 \end{pmatrix},$$

let

$$\begin{pmatrix} \dfrac{1}{\sqrt{2}} & \dfrac{1}{\sqrt{2}} \\ -\dfrac{1}{\sqrt{2}} & \dfrac{1}{\sqrt{2}} \end{pmatrix} \begin{pmatrix} x \\ y \end{pmatrix} = \begin{pmatrix} v_1 \\ v_2 \end{pmatrix},$$

then

$$\begin{pmatrix} v_1 \\ v_2 \end{pmatrix} A = \begin{pmatrix} \dfrac{1}{\sqrt{2}} & \dfrac{1}{\sqrt{2}} \\ -\dfrac{1}{\sqrt{2}} & \dfrac{1}{\sqrt{2}} \end{pmatrix} \begin{pmatrix} x \\ y \end{pmatrix} A$$

$$= \begin{pmatrix} \dfrac{1}{\sqrt{2}} & \dfrac{1}{\sqrt{2}} \\ -\dfrac{1}{\sqrt{2}} & \dfrac{1}{\sqrt{2}} \end{pmatrix} \begin{pmatrix} \lambda & 0 \\ 0 & \dfrac{1}{\lambda} \end{pmatrix} \begin{pmatrix} x \\ y \end{pmatrix}$$

$$= \begin{pmatrix} \dfrac{1}{\sqrt{2}} & \dfrac{1}{\sqrt{2}} \\ -\dfrac{1}{\sqrt{2}} & \dfrac{1}{\sqrt{2}} \end{pmatrix} \begin{pmatrix} \lambda & 0 \\ 0 & \dfrac{1}{\lambda} \end{pmatrix} \begin{pmatrix} \dfrac{1}{\sqrt{2}} & -\dfrac{1}{\sqrt{2}} \\ \dfrac{1}{\sqrt{2}} & \dfrac{1}{\sqrt{2}} \end{pmatrix} \begin{pmatrix} v_1 \\ v_2 \end{pmatrix}$$

$$= \begin{pmatrix} \dfrac{\lambda^{-1} + \lambda}{2} & \dfrac{\lambda^{-1} - \lambda}{2} \\ \dfrac{\lambda^{-1} - \lambda}{2} & \dfrac{\lambda^{-1} + \lambda}{2} \end{pmatrix} \begin{pmatrix} v_1 \\ v_2 \end{pmatrix}.$$

Now let $\frac{1}{2}(\lambda + \lambda^{-1}) = \cosh \psi$, then

$$\begin{pmatrix} v_1 \\ v_2 \end{pmatrix} A = \begin{pmatrix} \cosh \psi & \sinh \psi \\ \sinh \psi & \cosh \psi \end{pmatrix} \begin{pmatrix} v_1 \\ v_2 \end{pmatrix}$$

and

$$\begin{pmatrix} v_1 \\ v_2 \end{pmatrix} F \begin{pmatrix} v_1 \\ v_2 \end{pmatrix}' = \begin{pmatrix} 1 & 0 \\ 0 & -1 \end{pmatrix}.$$

Let T be a Lorentz matrix which takes v_1 and v_2 as its last two rows; then we have

$$TAT^{-1} = \begin{pmatrix} a_{11} & a_{12} & \cdots & a_{1m-2} & 0 & 0 \\ \cdots\cdots\cdots\cdots\cdots\cdots\cdots\cdots\cdots\cdots\cdots\cdots\cdots\cdots\cdots\cdots\cdots\cdots \\ a_{m-2,1} & a_{m-2,2} & \cdots & a_{m-2,m-2} & 0 & 0 \\ 0 & 0 & \cdots & 0 & \cosh \psi & \sinh \psi \\ 0 & 0 & \cdots & 0 & \sinh \psi & \cosh \psi \end{pmatrix},$$

where $(a_{ij})_{1 \le i, j \le m-2}$ is an orthogonal matrix.

(d) If A has a complex eigenvalue whose absolute value is equal to 1, then by the same method used in §3, it can be proved that there exists a Lorentz matrix T such that

$$TAT^{-1} = \begin{pmatrix} \cos\theta & \sin\theta & 0 & \cdots & 0 \\ -\sin\theta & \cos\theta & 0 & \cdots & 0 \\ 0 & 0 & a_{33} & \cdots & a_{3m} \\ \cdots\cdots\cdots\cdots\cdots\cdots\cdots \\ 0 & 0 & a_{m3} & \cdots & a_{mm} \end{pmatrix},$$

where $(a_{ij})_{3 \leqslant i, j \leqslant m}$ is an $(m-2) \times (m-2)$ Lorentz matrix.

4.6 Continuation

From the results of the preceding section, we know that the case that remains to be studied is the one where A has only the eigenvalues $+1$ and -1.

Suppose A has eigenvalue 1, and x is an eigenvector corresponding to A, i.e.

$$xA = x. \tag{1}$$

(a) Suppose there exists a vector satisfying (1) such that

$$xFx' > 0.$$

There is no harm in assuming that $xFx' = 1$. Choose a Lorentz matrix T whose first row is x, then

$$TAT^{-1} = \begin{pmatrix} 1 & 0 \\ 0 & A_1 \end{pmatrix},$$

where A_1 is an $(m-1) \times (m-1)$ Lorentz matrix.

(b) Suppose there exists a vector satisfying (1) such that

$$xFx' < 0.$$

Again we may assume that $xFx' = -1$. Choose a Lorentz matrix T which takes x to be its last row, then

$$TAT^{-1} = \begin{pmatrix} A_2 & 0 \\ 0 & -1 \end{pmatrix},$$

where A_2 is an $(m-1) \times (m-1)$ orthogonal matrix.

(c) The main difficulty lies in dealing with the case in which all x which satisfy $xA = x$, also satisfy $xFx' = 0$.

If there exist two linearly independent vectors x, y such that

$$\begin{aligned} xA &= x, & xFx' &= 0, \\ yA &= y, & yFy' &= 0, \end{aligned} \tag{2}$$

then since

$$(\lambda x + \mu y)A = \lambda x + \mu y,$$

we therefore also have

$$(\lambda x + \mu y)F(\lambda x + \mu y)' = 0,$$

i.e. we have

$$xFy' = 0,$$

and

$$\begin{pmatrix} x \\ y \end{pmatrix} F \begin{pmatrix} x \\ y \end{pmatrix}' = \begin{pmatrix} 0 & 0 \\ 0 & 0 \end{pmatrix}.$$

However from the result of §4 we know that this is impossible.

Therefore there exists only one vector x (up to a constant multiple) such that $xA = x$, $xFx' = 0$.

(d) If A has an eigenvalue equal to -1, then assume that

$$zA = -z.$$

If $zFz' \neq 0$, we may then solve the problem using the same methods as in (a) and (b). If $zFz' = 0$, then

$$xFz' = xAFA'z' = -xFz',$$

i.e. $xFz' = 0$. Furthermore

$$\begin{pmatrix} x \\ z \end{pmatrix} F \begin{pmatrix} x \\ z \end{pmatrix}' = \begin{pmatrix} 0 & 0 \\ 0 & 0 \end{pmatrix},$$

and this is impossible.

(e) From (c) and (d) we know that we now only need to investigate the case where A has only $+1$ (or -1) as an eigenvalue, and that there exists only one linearly independent eigenvector corresponding to $+1$. (That is to say,

$$\begin{pmatrix} 1 & 0 & 0 & \cdots \\ 1 & 1 & 0 & \cdots \\ 0 & 1 & 1 & \cdots \\ \cdots\cdots\cdots\cdots \end{pmatrix}$$

is its Jordan canonical form.) When $m \geq 2$, there would exist a vector y such that

$$yA = y + x. \tag{3}$$

From

$$yFy' = yAFA'y' = (x + y)F(x + y)' = yFy' + 2xFy',$$

we then obtain

$$xFy' = 0. \tag{4}$$

Since

$$\begin{pmatrix} x \\ y \end{pmatrix} F \begin{pmatrix} x \\ y \end{pmatrix}' = \begin{pmatrix} 0 & 0 \\ 0 & yFy' \end{pmatrix}, \tag{5}$$

thus $yFy' = a \neq 0$. If $m = 2$, then the left side of (5) is nonsingular and this is impossible. Therefore $m \geq 3$.

In other words, there would exist a vector z such that

$$zA = z + y. \tag{6}$$

From

$$zFy' = zAFA'y' = (z + y)F(y + x)' = zFy + yFy' + zFx',$$

we have

$$xFz' = -yFy' = -a, \tag{7}$$

and from

$$zFz' = zAFA'z' = (z + y)F(z + y)' = zFz' + 2yFz' + yFy',$$

we have

$$yFz' = -\tfrac{1}{2}yFy' = -\tfrac{1}{2}a.$$

Thus we can derive the following:

$$\begin{pmatrix} x \\ y \\ z \end{pmatrix} F \begin{pmatrix} x \\ y \\ z \end{pmatrix}' = \begin{pmatrix} 0 & 0 & -a \\ 0 & a & -\tfrac{1}{2}a \\ -a & -\tfrac{1}{2}a & b \end{pmatrix}, \qquad b = zFz'. \tag{8}$$

If there exists yet another vector w such that

$$wA = w + z, \tag{9}$$

then from

$$xFw' = xAFA'w' = xF(w + z)' = xFw' + xFz',$$

we have $xFz' = 0$. But this contradicts the fact that $a \neq 0$. Therefore $m = 3$, and

$$\begin{pmatrix} x \\ y \\ z \end{pmatrix} A = \begin{pmatrix} 1 & 0 & 0 \\ 1 & 1 & 0 \\ 0 & 1 & 1 \end{pmatrix} \begin{pmatrix} x \\ y \\ z \end{pmatrix}. \tag{10}$$

(f) Let us now consider the symmetric matrix (8). It has determinant $-a^3$ ($\neq 0$), and it is a quadratic form of type $(+1, +1, -1)$. Therefore $a > 0$. Let

$$x^* = \frac{1}{\sqrt{a}} x, \qquad y^* = \frac{1}{\sqrt{a}} y,$$

$$z^* = \frac{1}{\sqrt{a}} (\lambda x + z), \qquad \lambda = \frac{b}{2a}.$$

Then

$$x^*Fx^{*'} = 0, \qquad x^*Fy^{*'} = 0, \qquad x^*Fz^{*'} = -1$$
$$y^*Fy^* = 1, \qquad y^*Fz^* = -1/2,$$

and

$$z^*Fz^{*'} = \frac{1}{\sqrt{a}} (2\lambda xFz' + zFz') = \frac{1}{\sqrt{a}} (-2a\lambda + b) = 0.$$

On the other hand, we have as before,

$$\begin{pmatrix} x^* \\ y^* \\ z^* \end{pmatrix} A = \begin{pmatrix} 1 & 0 & 0 \\ 1 & 1 & 0 \\ 0 & 1 & 1 \end{pmatrix} \begin{pmatrix} x^* \\ y^* \\ z^* \end{pmatrix}.$$

Cancelling the asterisk (*), we may safely assume that our original a and b are now equal to 1 and 0, respectively. We therefore now have

$$\begin{pmatrix} x \\ y \\ z \end{pmatrix} F \begin{pmatrix} x \\ y \\ z \end{pmatrix}' = \begin{pmatrix} 0 & 0 & -1 \\ 0 & 1 & -\frac{1}{2} \\ -1 & -\frac{1}{2} & 0 \end{pmatrix} \tag{11}$$

and

$$\begin{pmatrix} x \\ y \\ z \end{pmatrix} A = \begin{pmatrix} 1 & 0 & 0 \\ 1 & 1 & 0 \\ 0 & 1 & 1 \end{pmatrix} \begin{pmatrix} x \\ y \\ z \end{pmatrix}.$$

It is now not difficult to directly prove, that by letting

$$P = \begin{pmatrix} \sqrt{2} & -\sqrt{2} & -2 \\ 0 & -\sqrt{2} & -1 \\ -\dfrac{\sqrt{2}}{8} & \dfrac{\sqrt{2}}{8} & -\dfrac{1}{4} \end{pmatrix},$$

then

$$P \begin{pmatrix} 0 & 3 & 2\sqrt{2} \\ -1 & 0 & 0 \\ 0 & 2\sqrt{2} & 3 \end{pmatrix} P^{-1} = \begin{pmatrix} 1 & 0 & 0 \\ 1 & 1 & 0 \\ 0 & 1 & 1 \end{pmatrix}$$

and

$$P \begin{pmatrix} 1 & 0 & 0 \\ 0 & 1 & 0 \\ 0 & 0 & -1 \end{pmatrix} P' = \begin{pmatrix} 0 & 0 & -1 \\ 0 & 1 & -\frac{1}{2} \\ -1 & -\frac{1}{2} & 0 \end{pmatrix}.$$

Let

$$P^{-1} \begin{pmatrix} x \\ y \\ z \end{pmatrix} = T,$$

then

$$TA = P^{-1} \begin{pmatrix} x \\ y \\ z \end{pmatrix} A = P^{-1} \begin{pmatrix} 1 & 0 & 0 \\ 1 & 1 & 0 \\ 0 & 1 & 1 \end{pmatrix} \begin{pmatrix} x \\ y \\ z \end{pmatrix}$$

$$= \begin{pmatrix} 0 & 3 & 2\sqrt{2} \\ -1 & 0 & 0 \\ 0 & 2\sqrt{2} & 3 \end{pmatrix} T,$$

and moreover,

$$TFT' = P^{-1} \begin{pmatrix} x \\ y \\ z \end{pmatrix} F \begin{pmatrix} x \\ y \\ z \end{pmatrix}' P'^{-1}$$

$$= P^{-1} \begin{pmatrix} 0 & 0 & -1 \\ 0 & 1 & -\frac{1}{2} \\ -1 & -\frac{1}{2} & 0 \end{pmatrix} P'^{-1} = F.$$

☐

(The same method applies in the case where the eigenvalue is -1.)

4.7 The Canonical Forms of Lorentz Similarity

So far, we know that any Lorentz matrix is similar to one of the following four types of expressions:

(a) $\begin{pmatrix} \cos\theta_1 & \sin\theta_1 \\ -\sin\theta_1 & \cos\theta_1 \end{pmatrix} + \cdots + \begin{pmatrix} \cos\theta_v & \sin\theta_v \\ -\sin\theta_v & \cos\theta_v \end{pmatrix}$

$+ \underbrace{1 + \cdots + 1}_{t} + \underbrace{(-1) + \cdots + (-1)}_{s} + (\pm 1),$

(b) $\begin{pmatrix} \cos\theta_1 & \sin\theta_1 \\ -\sin\theta_1 & \cos\theta_1 \end{pmatrix} + \cdots + \begin{pmatrix} \cos\theta_v & \sin\theta_v \\ -\sin\theta_v & \cos\theta_v \end{pmatrix}$

$+ \underbrace{1 + \cdots + 1}_{t} + \underbrace{(-1) + \cdots + (-1)}_{s}$

$+ \begin{pmatrix} 0 & 3 & 2\sqrt{2} \\ -1 & 0 & 0 \\ 0 & 2\sqrt{2} & 3 \end{pmatrix},$

(c) $\begin{pmatrix} \cos\theta_1 & \sin\theta_1 \\ -\sin\theta_1 & \cos\theta_1 \end{pmatrix} + \cdots + \begin{pmatrix} \cos\theta_v & \sin\theta_v \\ -\sin\theta_v & \cos\theta_v \end{pmatrix}$

$+ \underbrace{1 + \cdots + 1}_{t} + \underbrace{(-1) + \cdots + (-1)}_{s}$

$+ \begin{pmatrix} 0 & -3 & -2\sqrt{2} \\ 1 & 0 & 0 \\ 0 & -2\sqrt{2} & -3 \end{pmatrix},$

(d) $\begin{pmatrix} \cos\theta_1 & \sin\theta_1 \\ -\sin\theta_1 & \cos\theta_1 \end{pmatrix} + \cdots + \begin{pmatrix} \cos\theta_v & \sin\theta_v \\ -\sin\theta_v & \cos\theta_v \end{pmatrix}$

$+ \underbrace{1 + \cdots + 1}_{t} + \underbrace{(-1) + \cdots + (-1)}_{s}$

$+ \begin{pmatrix} \cosh\psi & \sinh\psi \\ \sinh\psi & \cosh\psi \end{pmatrix}.$

Note the difference when dealing instead with an orthogonal matrix: the similarity property among orthogonal matrices may be completely determined from the characteristic polynomial. However, that of a Lorentz matrix not only cannot be determined in this way, but it also cannot be completely determined from the elementary divisors. In addition to the elementary divisors, it would be necessary to look at the sign of the element in the lower right hand corner, i.e. it is fairly simple to prove that the necessary and sufficient conditions for Lorentz similarity of Lorentz matrices are that they have similar elementary divisors and that the elements of the lower right hand corner have the same sign.

It is also not difficult to prove that the elementary divisors of Lorentz matrices corresponding to eigenvalues not equal to ± 1 are all simple, and that when the eigenvalues are equal to ± 1, the corresponding elementary divisors have two possibilities: they are either of degree 1 or of degree 3.

4.8 Involution

Definition. If an $(n + 2) \times (n + 2)$ Lorentz matrix A satisfies

$$A^2 = \rho I, \qquad \rho = \bar\rho \neq 0, \tag{1}$$

then it is called an *involution* or an *involutory Lorentz matrix*.

Since $\det A^2 = 1$, therefore $\rho^{n+2} = 1$, and thus $\rho = 1$.

From the previous section, we know that an involutory Lorentz matrix is Lorentz similar to the canonical form

$$[1, \ldots, 1, -1, \ldots, -1, \pm 1], \tag{2}$$

so that the following two classes of involutory Lorentz matrices are of basic importance:

(i) the involutory Lorentz matrices which are Lorentz similar to

$$[-1, 1, \ldots, 1, 1, 1] \tag{3}$$

are called *mirror reflections*, or *spatial symmetries*,

(ii) the involutory Lorentz matrices which are Lorentz similar to

$$[1, \ldots, 1, 1, -1] \tag{4}$$

are called *inversions*, or *time symmetries*.

Their basic importance lies in that the other involutions can all be obtained as a finite product of those involutions belonging to (i) and (ii); at the same time, the involutions of (i) and (ii) are not mutually Lorentz similar.

Let us now determine the general formulas for reflection and inversion:

(i) Since with respect to any reflection A, there exists a Lorentz matrix P such that

$$PAP^{-1} = [-1, 1, \ldots, 1, 1, 1] = I - 2[1, 0, \ldots, 0],$$

we therefore have that

$$A = I - 2P^{-1}[1, 0, \ldots, 0]P.$$

Furthermore, from the fact that $P \in L_{(n+1,1)}$, we know that $P^{-1} = FP'F$. Substitution then gives

$$A = I - 2[1, \ldots, 1, -1]P'[1, \ldots, 1, -1][1, 0, \ldots, 0]P$$
$$= I - 2[1, \ldots, 1, -1]p'p,$$

wherein p is the first row vector of P. Since $P \in L_{(n+1,1)}$, therefore $pFp' = 1$, and so the general formula for reflection is

$$A = I - 2Fp'p, \qquad pFp' = 1. \tag{5}$$

(ii) Since with respect to any inversion A, there exists a Lorentz matrix P such that

$$PAP' = [1, \ldots, 1, -1] = I - 2[0, \ldots, 0, 1],$$

therefore,

$$A = I - 2P^{-1}[0, \ldots, 0, 1]P;$$

and since $P^{-1} = FP'F$, by substitution we know that

$$A = I + 2FP'[0, \ldots, 0, 1]P = I + 2Fq'q,$$

wherein q is the $(n+2)$th row vector of P. Since $P \in L_{(n+1,1)}$, therefore $qFq' = -1$, and hence the general formula for inversion is

$$A = I + 2Fq'q, \qquad qFq' = -1. \tag{6}$$

Applying (1.3) to any involutory Lorentz matrix results in the matrix

$$M = \left[I^{(n)}, \begin{pmatrix} 1 & -1 \\ 1 & 1 \end{pmatrix} \right]^{-1} A \left[I^{(n)}, \begin{pmatrix} 1 & -1 \\ 1 & 1 \end{pmatrix} \right]. \tag{7}$$

We shall also call the transformation

$$(y, yy', 1) = \rho(x, xx', 1)M \tag{8}$$

an *involution*. Likewise, when A is a reflection, the involution (8) is called a *reflection*, and when A is an inversion, the involution (8) is called an *inversion*.

The canonical form of a reflection is

$$y_1 = -x_1, \quad y_2 = x_2, \quad \ldots, \quad y_n = x_n \tag{9}$$

and that of an inversion is

$$y = -x/xx' \tag{10}$$

Since, for the former, we have

$$A = [-1, 1, \ldots, 1],$$

therefore

$$M = [-1, 1, \ldots, 1];$$

and since, for the latter, we have

$$A = [1, \ldots, 1, -1],$$

therefore

$$M = \left[1, \ldots, 1, \begin{pmatrix} 0 & -1 \\ -1 & 0 \end{pmatrix} \right].$$

Hence

$$y = \rho x, \quad yy' = -\rho, \quad -\rho xx' = 1,$$

i.e.

$$\rho = -1/xx',$$

and consequently,

$$y = -x/xx'.$$

CHAPTER 5

The Fundamental Theorem of Spherical Geometry—with a Discussion of the Fundamental Theorem of Special Relativity

5.1 Introduction

In 1946, when the author was studying the geometry of matrices, he used a method which can be used to deal with the fundamental theorem of n-dimensional spherical space; that is to say, from the property of the tangency of spheres one can derive the fundamental theorem of spherical geometry, so that neither the analycity nor even the continuity of certain transformations need ever be considered.

In this book we shall only discuss the spherical geometry of 3-dimensional space, the reason being that on the one hand the latter is relatively concrete and thus more easily understood, and on the other these results may thereby become more accessible to physicists. Actually, 3-dimensional spherical geometry is just special relativity in a different guise, but this point often goes unnoticed, as in the case of V. A. Fok, who in 1961 wrote the book entitled *The Theory of Space, Time and Gravitation*, (Second revised edition, English translation by N. Kemmer, Macmillan, New York, 1964). In this book the author still goes by the old method of using Riemannian geometry and the theory of Lie groups. Moreover the Chinese, English, German and other translations of this book also failed to take note of the aforementioned to provide the needed clarification.

Regarding the theory of special relativity, there were originally two assumptions:

(a) The principle of relativity requires that uniform linear motions remain uniform linear motions.
(b) The principle of the constancy of the speed of light assumes that light travels linearly and with constant speed c.

The method we shall adopt to deal with this situation is to use the principle of the constancy of the speed of light to deduce the Lorentz group; that is to say, "uniform motions remain uniform motions" as required by the principle of relativity is now a deduction and not an assumption. This has the advantage that if we wish to either verify or overthrow the above two assumptions, it would suffice to do so to the assumption of constant light speed alone. The generalization to n-dimensional spherical geometry and to the general geometry of Hermitian matrices will not be our main concern here.

Let the sphere under consideration have center (x, y, z) and radius R. If R is positive, then the sphere has an outward pointing arrowhead (Fig. 1); if R is negative, the sphere has an inward pointing arrowhead (Fig. 2).

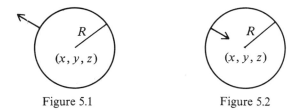

Figure 5.1   Figure 5.2

The condition for two tangent spheres is that the arrowheads point in the same direction at the point of tangency. The condition for two tangent spheres, (x, y, z, R), (x_1, y_1, z_1, R_1) is that

$$(x - x_1)^2 + (y - y_1)^2 + (z - z_1)^2 = (R - R_1)^2. \tag{1}$$

When R, R_1 have the same sign, the spheres are internally tangent, and when they have opposite signs, the spheres are externally tangent, i.e. as in Fig. 3. The spherical geometry under consideration then is the geometry of the space formed by taking the spheres as elements.

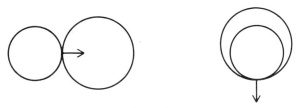

Figure 5.3

Let us use the 2×2 Hermitian matrix

$$H = \begin{pmatrix} R + x & y + iz \\ y - iz & R - x \end{pmatrix}$$

to express the sphere whose center is (x, y, z) and whose radius is R. Then the condition of tangency (1) is that the determinant of

$$H - H_1 = \begin{pmatrix} R - R_1 + x - x_1 & y - y_1 + i(z - z_1) \\ y - y_1 - i(z - z_1) & R - R_1 - (x - x_1) \end{pmatrix}$$

should be 0. In matrix geometry, this notion of tangency is what is called *coherence*. Our problem is to find the largest group of transformations from 2×2 Hermitian matrices into themselves which preserve the relation of coherence.

In the language of the theory of relativity, the event with spatial co-ordinates (x, y, z) and time coordinate t may be represented by the Hermitian matrix

$$\begin{pmatrix} ct + x & y + iz \\ y - iz & ct - x \end{pmatrix}.$$

The condition of coherence is that the spatial distance between two points is equal to the product of the time needed to traverse this distance and the speed of light.

Unlike Fok, we only need 2×2 matrices here.

5.2 Uniform Linear Motion

Since the occurrence of an event requires a time t and spatial coordinates (x, y, z), uniform linear motion may be expressed in the form:

$$x - x_0 = v_x(t - t_0), \qquad y - y_0 = v_y(t - t_0),$$
$$z - z_0 = v_z(t - t_0). \tag{1}$$

This is a uniform linear motion in the direction $v_x : v_y : v_z$, with speed $v = (v_x^2 + v_y^2 + v_z^2)^{1/2}$, and passing through (x_0, y_0, z_0) at time $t = t_0$.

Rewrite (1) as:

$$\begin{cases} x - x_0 = \alpha \tau, \\ y - y_0 = \beta \tau, \\ z - z_0 = \gamma \tau, \\ t - t_0 = \delta \tau. \end{cases} \qquad -\infty < \tau < \infty. \tag{2}$$

Thus it is possible to see the one-to-one correspondence between the uniform linear motion (1) of 3-dimensional space and the straight line (2) of 4-dimensional space.

From the fundamental theorem of affine geometry of 4-dimensional space (cf. Hua Luogeng (Hua Loo-keng), Wan Zhexian, *Classical Groups* (Chinese), Science Press, Shanghai, 1962), we know that the transformation which takes 4-dimensional space one-one into itself and which preserves linearity is indeed an affine transformation.

If we introduce infinitely distant points to extend affine space into projective space, then from the fundamental theorem of projective geometry (cf. *Classical Groups*, loc. cit.), we know that transformation thereby derived is a projective transformation. This is the first conclusion of Appendix I in Fok's book. Here, not only did we not require that the transformation possesses third order partial derivatives, but we did not even assume continuity.

5.3 The Geometry of Hermitian Matrices

Unless otherwise specified, capital letters $A, B, C, \ldots$ represent 2×2 complex matrices. The matrices which satisfy $P = \bar{P}'$ are called *Hermitian matrices*, or more simply H *matrices*. Clearly the transformation

$$X^* = AX\bar{A}' + X_0, \qquad \bar{X}'_0 = X_0 \tag{1}$$

takes a Hermitian matrix X into another Hermitian matrix X^*, where A is an invertible matrix. In addition, we also have the transformations

$$X^* = -X, \tag{2}$$

$$X^* = X'. \tag{3}$$

Consider two Hermitian matrices X_1, X_2; if the rank of $X_1 - X_2$ is equal to 1, then X_1, X_2 are said to be *coherent*. The reason for defining it in this way is to facilitate the generalization of the results of this section to $n \times n$ Hermitian matrices. So now we have

$$|X_1 - X_2| = 0. \tag{4}$$

Fundamental Theorem. *The transformations which give a one-one correspondence of the set of Hermitian matrices onto itself and which moreover preserve the relation of coherence are exactly the group of transformations generated by* (1), (2) *and* (3).

Postponing the proof to a later section (§10), we shall first clarify the relationship between the group generated by (1), (2) and (3) and the Lorentz group. Corresponding to the Hermitian matrix

$$X = \begin{pmatrix} ct + x & y + iz \\ y - iz & ct - x \end{pmatrix}, \tag{5}$$

there exists a vector

$$\omega = (ct, x, y, z). \tag{6}$$

To each transformation among Hermitian matrices X, there corresponds a transformation among vectors ω. Thus the transformation

$$X^* = X + X_0 \tag{7}$$

corresponds to the translation

$$\omega^* = \omega + \omega_0. \tag{8}$$

For this reason we need only consider the transformation

$$X^* = AX\bar{A}' \tag{9}$$

together with (2) and (3). The linear transformation to which they correspond is denoted by

$$\omega^* = \omega L(A). \tag{10}$$

Taking the determinant of (10) gives

$$\omega^*[1, -1, -1, -1]\omega^{*\prime} = \omega[1, -1, -1, -1]\omega'|A\bar{A}'|,$$

i.e.

$$\omega L[1, -1, -1, -1]L'\omega' = \omega[1, -1, -1, -1]\omega'|A\bar{A}'|.$$

Therefore

$$L[1, -1, -1, -1]L' = |A\bar{A}'|[1, -1, -1, -1], \tag{11}$$

i.e. $L(A)/|A\bar{A}'|^{1/2}$ is a Lorentz transformation. However, noting that A and $e^{i\theta}A$ represent the same Lorentz transformation, we may assume that

$$|A| = \rho^2 > 0,$$

so that when $A = \rho I$, the corresponding transformation is a *homothetic* transformation. Thus we may let $A = \rho B$ with $|B| = 1$. Moreover, $\pm A$ correspond to the same Lorentz transformation as well as to the same homothetic transformation. Henceforth we shall suppose that $|A| = 1$.

Let us first study some of the special relationships between A and $L(A)$:

(i) $\qquad A = \begin{pmatrix} \cos \frac{1}{2}\theta, & -\sin \frac{1}{2}\theta \\ \sin \frac{1}{2}\theta, & \cos \frac{1}{2}\theta \end{pmatrix}$,

$$L(A) = \begin{pmatrix} 1 & 0 & 0 & 0 \\ 0 & \cos \theta & \sin \theta & 0 \\ 0 & -\sin \theta & \cos \theta & 0 \\ 0 & 0 & 0 & 1 \end{pmatrix},$$

i.e. a rotation around the z-axis.

(ii) $\qquad A = \begin{pmatrix} e^{i\theta/2} & 0 \\ 0 & e^{-i\theta/2} \end{pmatrix}$,

$$L(A) = \begin{pmatrix} 1 & 0 & 0 & 0 \\ 0 & 1 & 0 & 0 \\ 0 & 0 & \cos \theta & \sin \theta \\ 0 & 0 & -\sin \theta & \cos \theta \end{pmatrix},$$

i.e. a rotation around the x-axis.

(iii) $A = \begin{pmatrix} \cos\frac{1}{2}\theta, & -i\sin\frac{1}{2}\theta \\ -i\sin\frac{1}{2}\theta, & \cos\frac{1}{2}\theta \end{pmatrix}$,

$$L(A) = \begin{pmatrix} 1 & 0 & 0 & 0 \\ 0 & \cos\theta & 0 & \sin\theta \\ 0 & 0 & 1 & 0 \\ 0 & -\sin\theta & 0 & \cos\theta \end{pmatrix},$$

i.e. a rotation around the y-axis.

(iv) $A = \begin{pmatrix} e^{\psi/2} & 0 \\ 0 & e^{-\psi/2} \end{pmatrix} (\psi > 0)$,

$$L(A) = \begin{pmatrix} \cosh\psi & \sinh\psi & 0 & 0 \\ \sinh\psi & \cosh\psi & 0 & 0 \\ 0 & 0 & 1 & 0 \\ 0 & 0 & 0 & 1 \end{pmatrix},$$

which is a hyperbolic rotation.

(v) $X^* = X'$, $t^* = t$, $x^* = x$, $y^* = y$, $z^* = -z$,

and this is just a spatial reflection.

(vi) $X^* = -AX'\bar{A}'$, $A = \begin{pmatrix} 0 & -1 \\ 1 & 0 \end{pmatrix}$;

$t^* = -t$, $x^* = x$, $y^* = y$, $z^* = z$,

and this is a time inversion.

From (i), (ii) and (iii), we see that $L(A)$ generates the whole rotation group. Moreover, A generates the 2-dimensional *special unitary group*, i.e. the 2×2 unitary matrices with determinant 1. Thus there is a one-one correspondence between the special unitary group, modulo ± 1, with the rotation group. Whereas the $L(A)$'s of (i), (ii), (iii) and (v) generate the orthogonal group, the $L(A)$'s of (i)–(iv) generate all the transformations in $L_+^\uparrow$. The $L(A)$'s of (i)–(v) give all the transformations in the Lorentz group.

Incidentally, this justifies our choices of signs at various stages of the preceding development since we have now obtained, via $L(A)$, the whole Lorentz group of special relativity.

Remark. For any A ($|A| = 1$), there always exists a square matrix P ($|P| = 1$), such that

$$PAP^{-1} = \begin{pmatrix} \lambda & 0 \\ 0 & \lambda^{-1} \end{pmatrix}, \qquad \begin{pmatrix} 1 & 1 \\ 0 & 1 \end{pmatrix}.$$

These lead to the canonical forms of Lorentz matrices under Lorentzian similarity. The former gives rise to the so-called "Lorentz 4 screw". The latter is just the singular canonical form of §6 in Chapter 4, i.e. the so-called "null rotation". However in the case of 2×2 matrices, the fact is very clearly brought out that this singular canonical form corresponds to the case with a non-simple elementary divisor. To this day, the special relativistic significance of this type of transformation has remained a mystery to the author, however.

The physical significance of our fundamental theorem is that the fact that "light travels linearly with constant speed" characterizes Einstein's special relativity. Actually, if we assume that the speed of the passing of information has an upper bound, then this set of mathematical tools is still usable. From now on we shall use x_0, x_1, x_2, x_3 in place of t, x, y, z, respectively.

Additional Remark. In a seminar at the Chinese University of Science and Technology in the 1950's, the author already pointed out the applications of this fundamental theorem to the theory of special relativity. Recently, while on a lecture tour in England, the author was informed by Prof. Gunsen of the University of Birmingham of the work of Prof. E. C. Zeemann (see J. Math. Phys. 5 (1964), pp. 490–493). Under the assumptions that the speed of light remains constant and that the law of causality is valid (the latter means that the inequality between two spacetime points

$$ct^* - ct > ((x^* - x)^2 + (y^* - y)^2 + (z^* - z)^2)^{1/2}$$

should hold, or equivalently, that $H^* - H > 0$), Zeeman proved that the group of spacetime transformations is just the causality group. The causality group is generated by three types of transformations: (i) orthochronous Lorentz groups, i.e. linear self-mappings of spacetimes which preserve spacetime distances and time orientations, but may reverse spatial orientation, (ii) translation and (iii) dilation. In other words, this is the group of mappings $X \to X^*$ such that

$$X^* = \rho A X \bar{A}' + B, \qquad |A| = 1, \qquad \bar{B}' = B, \qquad \rho > 0,$$
$$X^* = X'.$$

This result is obvious from our fundamental theorem, for $X^* = -X$ can cause time reversal (cf. §11).

5.4 Affine Transformations Which Leave Invariant the Unit Sphere in 3-Dimensional Space

Theorem 1. *An affine transformation[1] which, when applied to the unit sphere leaves it invariant, is necessarily an orthogonal transformation (cf. Chapter 4, §3).*

[1] *Translator's note:* In this book, an *affine transformation* of a vector space V is a mapping $v \to f(v) + w$, where f is a *nonsingularity* linear map of V onto V and w is a fixed vector in V.

PROOF. Let us assume that

$$y = xA + b \tag{1}$$

is such an affine transformation, where x, y, b are 3-dimensional vectors and A is a 3×3 nonsingular matrix. (1) then transforms $xx' = 1$ into $yy' = 1$.

There exist orthogonal matrices Γ_1 and Γ_2 such that

$$\Gamma_1 A \Gamma_2 = \begin{pmatrix} \lambda_1 & 0 & 0 \\ 0 & \lambda_2 & 0 \\ 0 & 0 & \lambda_3 \end{pmatrix}, \qquad \lambda_r > 0.$$

(use the polar decomposition of A, for instance). Thus without loss of generality, we may assume

$$A = \begin{pmatrix} \lambda_1 & 0 & 0 \\ 0 & \lambda_2 & 0 \\ 0 & 0 & \lambda_3 \end{pmatrix},$$

where $\lambda_i > 0$. Letting x be the point

$$x = \left(\frac{1 - t^2}{1 + t^2}, \frac{2t}{1 + t^2}, 0 \right)$$

on the unit sphere, we obtain

$$1 = y_1^2 + y_2^2 + y_3^2 = \left(\lambda_1 \frac{1 - t^2}{1 + t^2} + b_1 \right)^2 + \left(\frac{2t\lambda_2}{1 + t^2} + b_2 \right)^2 + b_3^2,$$

so that

$$(1 + t^2)^2 = (\lambda_1(1 - t^2) + b_1(1 + t^2))^2 + (2t\lambda_2 + b_2(1 + t^2))^2 + b_3^2(1 + t^2)^2.$$

Comparing the coefficients of t, we obtain $\lambda_2 b_2 = 0$, which implies that $b_2 = 0$. Similarly, $b_1 = b_3 = 0$. We now compare the coefficients of t^4 and obtain $\lambda_1^2 = 1$, whence $\lambda_1 = 1$. Analagously we obtain $\lambda_2 = \lambda_3 = 1$. In other words, (1) is the identity transformation. It follows that $b = 0$. $\qquad \square$

For the convenience of future applications, we shall now write 3-dimensional space as well as the group of rigid motions as 2×2 matrices, that is, we shall rewrite the point

$$x = (x_1, x_2, x_3) \quad \text{as} \quad X = \begin{pmatrix} x_1 & x_2 + ix_3 \\ x_2 - ix_3 & -x_1 \end{pmatrix},$$

that is to say, the point (x_1, x_2, x_3) of 3-dimensional space is now expressed as an H matrix with $t = 0$.

The line

$$x = a + \lambda b$$

may be rewritten as

$$X = A + \lambda B \qquad (-\infty < \lambda < \infty),$$

the plane
$$ax' = \mu$$
as
$$\operatorname{tr}(AX) = 2\mu,$$
and the distance between the two points
$$(x - u)(x - u)'$$
as the determinant
$$-|X - U|.$$

From the preceding paragraph we now know that the group of rigid motions
$$y = x\Gamma + b$$
$$\Gamma\Gamma' = \mathrm{I}, \qquad |\Gamma| = 1$$
can be rewritten as
$$Y = UX\bar{U}' + B, \qquad \bar{B}' = B,$$
$$U\bar{U}' = \mathrm{I}, \qquad |U| = 1.$$
Note that the U on the right hand side of the equation and $-U$ represent the same transformation.

The *Euclidean group* is the set of all rigid motions with reflections adjoined, that is,
$$y_1 = x_1, \qquad y_2 = x_2, \qquad y_3 = -x_3,$$
or equivalently
$$Y = X'.$$
Hence the Euclidean group may be expressed as
$$Y = UX\bar{U}' + B.$$
with $Y = X'$.

5.5 Coherent Subspaces

We shall now return to the problem raised in §3. The space under discussion is the 4-dimensional spacetime; in other words, the space derived from all Hermitian matrices. The points in this space then just refer to a 2×2 Hermitian matrix, and the transformations refer to the affine transformation defined in §3. We shall use coherence relations to define several geometric configurations.

Definition. Let A, B be two coherent points. Then the set of all points which are contiguous to A, B is called a *coherent subspace*.

Theorem 1. *Any coherent subspace may be transformed to the canonical form,*

$$\begin{pmatrix} a & 0 \\ 0 & 0 \end{pmatrix}, \qquad -\infty < a < \infty. \tag{1}$$

That is to say, under affine transformations, the coherent subspaces become a transitive set.

PROOF. Under an affine transformation we may safely assume that

$$A = \begin{pmatrix} 0 & 0 \\ 0 & 0 \end{pmatrix}, \qquad B = \begin{pmatrix} 1 & 0 \\ 0 & 0 \end{pmatrix}.$$

Let

$$X = \begin{pmatrix} \alpha & \beta \\ \bar{\beta} & \gamma \end{pmatrix},$$

where α, γ are real. From the coherence relation,

$$|A - X| = |B - X| = 0,$$

we get

$$\alpha\gamma - |\beta|^2 = 0, \qquad (\alpha - 1)\gamma - |\beta|^2 = 0,$$

whence $\gamma = 0$, $\beta = 0$. $\qquad\qquad\qquad\qquad\qquad\qquad\qquad$ $\square$

Theorem 2. *Two distinct coherent subspaces Σ_1, and Σ_2 have at most one point in common; if they do have a point in common, then they can be simultaneously transformed into*

$$\Sigma_1: \begin{pmatrix} a & 0 \\ 0 & 0 \end{pmatrix}, \qquad -\infty < a < \infty,$$

$$\Sigma_2: \begin{pmatrix} 0 & 0 \\ 0 & b \end{pmatrix}, \qquad -\infty < b < \infty.$$

PROOF. From the definition we already know that distinct coherent subspaces cannot have two points in common.

Let us assume that the common point is 0. Furthermore we may assume that one of the subspaces is already in canonical form, say Σ_1. Then an element Σ_2 is necessarily of the form

$$d\begin{pmatrix} |s|^2 & s\bar{t} \\ \bar{s}t & |t|^2 \end{pmatrix}, \qquad t \neq 0, \qquad d \neq 0. \tag{2}$$

Now under the transformation

$$Y = \begin{pmatrix} 1 & -st^{-1} \\ 0 & 1 \end{pmatrix} X \begin{pmatrix} 1 & -st^{-1} \\ 0 & 1 \end{pmatrix}'$$

Σ_1 remains invariant, but (2) becomes

$$d\begin{pmatrix} 1 & -st^{-1} \\ 0 & 1 \end{pmatrix}\begin{pmatrix} s\bar{s} & s\bar{t} \\ \bar{s}t & t\bar{t} \end{pmatrix}\overline{\begin{pmatrix} 1 & -st^{-1} \\ 0 & 1 \end{pmatrix}}' = d\begin{pmatrix} 0 & 0 \\ 0 & |t|^2 \end{pmatrix}.$$

This, together with 0 give us Σ_2. □

Theorem 3. *Three coherent subspaces which have at most one common point may be simultaneously transformed into*

$$\Sigma_1: \quad \begin{pmatrix} a & 0 \\ 0 & 0 \end{pmatrix}, \qquad -\infty < a < \infty,$$

$$\Sigma_2: \quad \begin{pmatrix} 0 & 0 \\ 0 & b \end{pmatrix}, \qquad -\infty < b < \infty,$$

$$\Sigma_3: \quad d\begin{pmatrix} 1 & 1 \\ 1 & 1 \end{pmatrix}, \qquad -\infty < d < \infty.$$

PROOF. Suppose two of the subspaces, say Σ_1, Σ_2, are already in canonical form. Then Σ_3 contains an element in the form of (2), where neither s nor t may equal 0, or else Σ_3 would belong either to Σ_1 or Σ_2.

Under the transformation

$$Y = \begin{pmatrix} s^{-1} & 0 \\ 0 & t^{-1} \end{pmatrix} X \overline{\begin{pmatrix} s^{-1} & 0 \\ 0 & t^{-1} \end{pmatrix}}'$$

Σ_1 and Σ_2 remain invariant, but (2) becomes

$$d\begin{pmatrix} s^{-1} & 0 \\ 0 & t^{-1} \end{pmatrix}\begin{pmatrix} |s|^2 & s\bar{t} \\ \bar{s}t & |t|^2 \end{pmatrix}\overline{\begin{pmatrix} s^{-1} & 0 \\ 0 & t^{-1} \end{pmatrix}}' = d\begin{pmatrix} 1 & 1 \\ 1 & 1 \end{pmatrix}.$$ □

5.6 Phase Planes (or 2-Dimensional Phase Subspaces)

Definition. Let Σ_1, Σ_2 be two coherent subspaces, having at most one point in common. Then the set consisting of all points contiguous neither to Σ_1 nor to Σ_2 is said to be a *2-dimensional phase subspace* (or *phase plane*).

Theorem 1. *A phase plane is transitive, and its canonical form is*

$$\begin{pmatrix} 0 & \beta \\ \bar{\beta} & 0 \end{pmatrix}, \tag{1}$$

where β is any complex number.

PROOF. We may safely assume that Σ_1, Σ_2 are of the same form as in theorem 2 of §5. Let

$$X = \begin{pmatrix} \alpha & \beta \\ \bar{\beta} & \gamma \end{pmatrix}$$

be a point of the phase plane. Now, since there do not exist an a, b which satisfy

$$\begin{vmatrix} \alpha - a & \beta \\ \bar{\beta} & \gamma \end{vmatrix} = 0, \qquad \begin{vmatrix} \alpha & \beta \\ \bar{\beta} & \gamma - b \end{vmatrix} = 0,$$

therefore $\gamma = \alpha = 0$.   □

The general form of a 2-dimensional phase subspace is

$$X = A \begin{pmatrix} 0 & \tau \\ \bar{\tau} & 0 \end{pmatrix} \bar{A}' + B, \tag{2}$$

where τ is any complex number.

5.7 Phase Lines

Definition. If two distinct phase planes have more than one point of intersection, then the set of all such intersection points is called a *phase line* (or 1-*dimensional phase subspace*).

Theorem 1. *A phase line has the following canonical form*:

$$\begin{pmatrix} 0 & \rho \\ \rho & 0 \end{pmatrix}, \qquad -\infty < \rho < \infty. \tag{1}$$

PROOF. Assume that one of the phase planes is

$$\begin{pmatrix} 0 & \xi \\ \bar{\xi} & 0 \end{pmatrix}, \tag{2}$$

where ξ is a complex number. The other one is given by (2) of $\xi 6$. Their two points of intersection are then $\xi = \xi_0, \xi = \xi_1$. The affine transformation

$$Y = \begin{pmatrix} e^{i\theta} & 0 \\ 0 & e^{-i\theta} \end{pmatrix} \left(X - \begin{pmatrix} 0 & \xi_0 \\ \bar{\xi}_0 & 0 \end{pmatrix} \right) \begin{pmatrix} e^{-i\theta} & 0 \\ 0 & e^{i\theta} \end{pmatrix}$$

takes ξ_0, ξ_1 into $\xi = 0$ and $\xi = \rho_1$, respectively. (ρ_1 is a real number, and $\xi_1 - \xi_0 = \rho_1 e^{-2i\theta}$). From (2) of $\xi 6$, we have

$$0 = A \begin{pmatrix} 0 & \tau_0 \\ \bar{\tau}_0 & 0 \end{pmatrix} \bar{A}' + B, \tag{3}$$

$$\begin{pmatrix} 0 & \rho_1 \\ \rho_1 & 0 \end{pmatrix} = A \begin{pmatrix} 0 & \tau_1 \\ \bar{\tau}_1 & 0 \end{pmatrix} \bar{A}' + B. \tag{4}$$

Now, for any real number ρ, if we multiply (3) by $(1 - \rho/\rho_1)$ and add to it the product of (4) and ρ/ρ_1, we obtain

$$\begin{pmatrix} 0 & \rho \\ \rho & 0 \end{pmatrix} = A \begin{pmatrix} 0 & \left(1 - \dfrac{\rho}{\rho_1}\right)\tau_0 + \dfrac{\rho}{\rho_1}\tau_1 \\ \left(1 - \dfrac{\rho}{\rho_1}\right)\tau_0 + \dfrac{\rho}{\rho_1}\tau_1 & 0 \end{pmatrix} \bar{A}' + B,$$

that is, (1) is part of the intersection of the two phase planes.   □

Let us now prove that there are no points other than those in the intersection (1). If there exists another ξ_0 (a nonreal number), i.e.

$$\begin{pmatrix} 0 & \xi_0 \\ \xi_0 & 0 \end{pmatrix} = A \begin{pmatrix} 0 & \tau_2 \\ \bar{\tau}_2 & B \end{pmatrix} \bar{A}' + B, \qquad (5)$$

then for any complex number ξ, we may choose a real number β, such that $\xi - \beta\xi_0 = \alpha$ is real. (3) multiplied by $1 - \alpha - \beta$, added to (4) multiplied by α/ρ_1, and this sum added to (5) multiplied by β yields

$$\begin{pmatrix} 0 & \xi \\ \bar{\xi} & 0 \end{pmatrix} = A \begin{pmatrix} 0 & (1 - \alpha - \beta)\tau_0 + \dfrac{\alpha}{\rho_1}\tau_1 + \beta\tau_2 \\ (1 - \alpha - \beta)\tau_0 + \dfrac{\alpha}{\rho_1}\tau_1 + \beta\tau_2 & 0 \end{pmatrix} \bar{A}' + B,$$

so that this phase plane would be identical to phase plane (6.2).   □

5.8 Point Pairs

Depending on whether $|A - B| > 0$, $|A - B| = 0$ or $|A - B| < 0$, we classify correspondingly the point pair A, B into three types: (a) *causally related pair* (or *hyperbolic pair*), (b) *coherent pair* (or *parabolic pair*) and (c) *causally nonrelated pair* (or *elliptic pair*).

Theorem 1. *Any two arbitrary points on a phase plane are an elliptic pair.*

Theorem 2. *Under an affine transformation, any hyperbolic pair may be transformed into the canonical forms*

$$\begin{pmatrix} 0 & 0 \\ 0 & 0 \end{pmatrix}, \quad \begin{pmatrix} 1 & 0 \\ 0 & 1 \end{pmatrix};$$

any parabolic pair may be transformed into the canonical forms

$$\begin{pmatrix} 0 & 0 \\ 0 & 0 \end{pmatrix}, \quad \begin{pmatrix} 1 & 0 \\ 0 & 0 \end{pmatrix};$$

and any elliptic pair may be transformed into the canonical forms

$$\begin{pmatrix} 0 & 0 \\ 0 & 0 \end{pmatrix}, \quad \begin{pmatrix} 0 & 1 \\ 1 & 0 \end{pmatrix}.$$

Since the proof of theorem 2 is obvious, it is omitted here. From theorem 2 we can then establish theorem 1, as follows. Since any two points which have a causal relation may, simultaneously, be transformed into $\begin{pmatrix} 0 & 0 \\ 0 & 0 \end{pmatrix}$, $\begin{pmatrix} 1 & 0 \\ 0 & 1 \end{pmatrix}$ if they lie on the same phase plane, that is, if we have

$$0 = A \begin{pmatrix} 0 & \xi \\ \bar{\xi} & 0 \end{pmatrix} \bar{A}' + B, \qquad I = A \begin{pmatrix} 0 & \xi_1 \\ \bar{\xi}_1 & 0 \end{pmatrix} \bar{A}' + B,$$

then, subtracting and taking the determinant would give $1 = -|\xi - \xi_1|^2$, which is impossible. $\qquad\qquad\qquad\qquad\qquad\qquad\qquad\qquad\qquad\qquad$ □

Theorem 3. *Under a coherence preserving one-one correspondence, the type of any point pair remains invariant.*

PROOF. Since a parabolic pair is clearly transformed into same, we only need to prove that any elliptic pair is transformed into same. We may take the canonical form of the elliptic pair to be

$$\begin{pmatrix} 0 & 0 \\ 0 & 0 \end{pmatrix}, \quad \begin{pmatrix} 0 & 1 \\ 1 & 0 \end{pmatrix}.$$

The line joining the two points is just the phase line

$$\begin{pmatrix} 0 & \rho_1 \\ \rho_1 & 0 \end{pmatrix}, \qquad -\infty < \rho_1 < \infty.$$

Since the transformation takes phase planes into phase planes, it therefore also takes phase lines into phase lines. Furthermore, we determined in theorem 1 of §7 the general form of the phase line to be

$$A \begin{pmatrix} 0 & \rho \\ \rho & 0 \end{pmatrix} \bar{A}' + B, \qquad -\infty < \rho < \infty.$$

On this line any two points obviously constitute an elliptic pair. Hence the transformation takes an elliptic pair into an elliptic pair. It follows that the transformation also takes a hyperbolic pair into a hyperbolic pair. $\qquad$ □

5.9 3-Dimensional Phase Subspaces

Definition 1. Outside a phase plane S_2, there exists a point P which is not causally related to any point in S_2. The set of points on all the lines connecting such a point P to S_2 is called a 3-*dimensional phase subspace*.

Theorem 1. *The canonical form of a 3-dimensional phase subspace is*

$$\begin{pmatrix} x_1 & x_2 + ix_3 \\ x_2 - ix_3 & -x_1 \end{pmatrix}, \qquad -\infty < x_1, x_2, x_3 < \infty. \tag{1}$$

PROOF. Let us assume the phase plane S_2 to be

$$\begin{pmatrix} 0 & q \\ \bar{q} & 0 \end{pmatrix} \tag{2}$$

Let $P = \begin{pmatrix} p_0 & q_0 \\ \bar{q}_0 & r_0 \end{pmatrix}$ be a point outside S_2. Then under the transformation

$$Y = X - \begin{pmatrix} 0 & q \\ \bar{q}_0 & 0 \end{pmatrix}$$

(2) remains invariant, but the point P becomes

$$P \quad \begin{pmatrix} p_0 & 0 \\ 0 & r_0 \end{pmatrix}.$$

Now, since P is not causally related to 0, we therefore have $p_0 r_0 < 0$. Again using the transformation

$$\begin{pmatrix} \lambda & 0 \\ 0 & \lambda^{-1} \end{pmatrix} \begin{pmatrix} p_0 & 0 \\ 0 & r_0 \end{pmatrix} \begin{pmatrix} \lambda & 0 \\ 0 & \lambda^{-1} \end{pmatrix},$$

we may assume that

$$P = p \begin{pmatrix} 1 & 0 \\ 0 & -1 \end{pmatrix}. \tag{3}$$

The line joining the points (2) and (3) is

$$\mu p \begin{pmatrix} 1 & 0 \\ 0 & -1 \end{pmatrix} + (1 - \mu) \begin{pmatrix} 0 & q \\ \bar{q} & 0 \end{pmatrix} = \begin{pmatrix} \mu p & (1 - \mu)q \\ (1 - \mu)\bar{q} & -\mu p \end{pmatrix},$$

which then yields our desired result. □

5.10 Proof of the Fundamental Theorem

(a) Let π_3 be a 3-dimensional phase subspace and let P be a point outside π_3. Then there exists an affine transformation which takes π_3 into

$$\begin{pmatrix} x_1 & x_2 + ix_3 \\ x_2 - ix_3 & -x_1 \end{pmatrix} \tag{1}$$

and P into

$$I = \begin{pmatrix} 1 & 0 \\ 0 & 1 \end{pmatrix}. \tag{2}$$

The proof of this is as follows: assume that π_3 has already been transformed into (1). Letting

$$P = \begin{pmatrix} p_0 + p_1 & p_2 + ip_3 \\ p_2 - ip_3 & p_0 - p_1 \end{pmatrix}, \qquad p_0 \neq 0,$$

we then have the transformation

$$Y = \frac{1}{p_0}\left[X - \begin{pmatrix} p_1 & p_2 + ip_3 \\ p_2 - ip_3 & -p_1 \end{pmatrix} \right],$$

which takes (1) into itself and P into (2).

(b) Let

$$Y = \Phi(X) \tag{3}$$

be a one-one correspondence taking the set of Hermitian matrices onto itself and preserving the coherence relation. From the definition of 3-dimensional phase subspace and from §8 (note that the results there were obtained purely from the coherence relation), we may as well assume that

$$\Phi\begin{pmatrix} x_1 & x_2 + ix_3 \\ x_2 - ix_3 & -x_1 \end{pmatrix} = \begin{pmatrix} y_1 & y_2 + iy_3 \\ y_2 - iy_3 & -y_1 \end{pmatrix}.$$

This type of subspace is represented by π_3, and

$$\Phi(I) = I. \tag{4}$$

In π_3 we have an ordinary 3-dimensional space (x_1, x_2, x_3). It is not difficult to prove that the planes contained therein are 2-dimensional phase planes, and that lines contained therein are 1-dimensional phase subspaces. From the fundamental theorem of affine transformations, the transformations which take lines into lines are just affine transformations of 3-dimensional space. Furthermore, as a result of the coherence relation

$$\left| I - \begin{pmatrix} x_1 & x_2 + ix_3 \\ x_2 - ix_3 & -x_1 \end{pmatrix} \right| = 1 - (x_1^2 + x_2^2 + x_3^2) = 0,$$

such affine transformations must also leave invariant the unit sphere.

Thus from the results of §3 and §4 we know that when $X, Y \in \pi_3$ and $Y = \Phi(X)$, we have

$$Y = UX\bar{U}', \qquad U\bar{U}' = I.$$

(c) For this reason let us assume that

$$\Phi\begin{pmatrix} \xi_1 & \xi_2 + i\xi_3 \\ \xi_2 - i\xi_3 & -\xi_1 \end{pmatrix} = \begin{pmatrix} \xi_1 & \xi_2 + i\xi_3 \\ \xi_2 - i\xi_3 & -\xi_1 \end{pmatrix}. \tag{5}$$

The condition of coherence for the general point

$$\begin{pmatrix} x_0 + x_1 & x_2 + ix_3 \\ x_2 - ix_3 & x_0 - x_1 \end{pmatrix}, \qquad x_0 \neq 0,$$

and points on π_3 is

$$x_0^2 = (x_1 - \xi_1)^2 + (x_2 - \xi_2)^2 + (x_3 - \xi_3)^2. \tag{6}$$

From (3) we know that all ξ_1, ξ_2, ξ_3 satisfying (6) must also satisfy

$$y_0^2 = (y_1 - \xi_1)^2 + (y_2 - \xi_2)^2 + (y_3 - \xi_3)^2; \tag{7}$$

moreover the converse is true.

Two points on π_3 which satisfy (6) are $\xi_1 = x_1 \pm x_0$, $\xi_2 = x_2$, $\xi_3 = x_3$. Substituting into (7) we have

$$y_0^2 = (y_1 - x_1 \pm x_0)^2 + (y_2 - x_2)^2 + (y_3 - x_3)^2. \tag{8}$$

Since the formula is true for both the positive and negative signs, we thus have

$$(y_1 - x_1)x_0 = 0,$$

that is, $y_1 = x_1$. The same method shows that $x_2 = y_2$, $x_3 = y_3$ and $x_0 = \pm y_0$. Hence we have arrived at the conclusion that

$$\Phi \begin{pmatrix} x_0 + x_1 & x_2 + ix_3 \\ x_2 - ix_3 & x_0 - x_1 \end{pmatrix} = \begin{pmatrix} \pm x_0 + x_1 & x_2 + ix_3 \\ x_2 - ix_3 & \pm x_0 - x_1 \end{pmatrix}, \tag{9}$$

and the only difference between this and the identity transformation is in the sign.

(d) We already know that $\Phi(I) = I$. We shall prove in addition that under Φ,

$$\begin{pmatrix} 1 + \lambda & 0 \\ 0 & 1 \end{pmatrix} = \begin{pmatrix} 1 + \dfrac{\lambda}{2} + \dfrac{\lambda}{2} & 0 \\ 0 & 1 + \dfrac{\lambda}{2} - \dfrac{\lambda}{2} \end{pmatrix} \tag{10}$$

remains invariant. From (9) we know that if it is not invariant, then it can only be taken into

$$\begin{pmatrix} -\left(1 + \dfrac{\lambda}{2}\right) + \dfrac{\lambda}{2} & 0 \\ 0 & -\left(1 + \dfrac{\lambda}{2}\right) - \dfrac{\lambda}{2} \end{pmatrix} = \begin{pmatrix} -1 & 0 \\ 0 & -1 - \lambda \end{pmatrix},$$

and this is not coherent with I, which is a contradiction. We can use the same method to prove that

$$\begin{pmatrix} 1 & 0 \\ 0 & 1 + \mu \end{pmatrix} \tag{11}$$

is also invariant under Φ.

Let us now prove that under Φ,

$$\begin{pmatrix} 1 + \lambda & 0 \\ 0 & 1 + \mu \end{pmatrix} = \begin{pmatrix} 1 + \tfrac{1}{2}(\lambda + \mu) + \tfrac{1}{2}(\lambda - \mu) & 0 \\ 0 & 1 + \tfrac{1}{2}(\lambda + \mu) - \tfrac{1}{2}(\lambda - \mu) \end{pmatrix} \tag{12}$$

remains invariant. Suppose it is not invariant under Φ. Then again by (9), it can only be taken into

$$\begin{pmatrix} -1 - \frac{1}{2}(\lambda + \mu) + \frac{1}{2}(\lambda - \mu) & 0 \\ 0 & -1 - \frac{1}{2}(\lambda + \mu) - \frac{1}{2}(\lambda - \mu) \end{pmatrix} = \begin{pmatrix} -1 - \mu & 0 \\ 0 & -1 - \lambda \end{pmatrix},$$

and this point can be coherent with (10) and (11), respectively, only if

$$(2 + \lambda + \mu)(2 + \lambda) = 0, \qquad (2 + \lambda + \mu)(2 + \mu) = 0,$$

that is, only if $\lambda = \mu = -2$ and $\lambda + \mu = -2$, that is, only if the point is one of two possibilities:

$$\begin{pmatrix} -1 & 0 \\ 0 & -1 \end{pmatrix}, \quad \text{or} \quad \begin{pmatrix} 1 + \lambda & 0 \\ 0 & -(1 + \lambda) \end{pmatrix}.$$

In the case of the latter, it is already invariant; in the case of the former, it is transformed either into itself or into I, which is impossible. So we have invariance in any case.

(e) We now already know that a square matrix of the form

$$\begin{pmatrix} \lambda & 0 \\ 0 & \mu \end{pmatrix} = \begin{pmatrix} y_0 + y_1 & 0 \\ 0 & y_0 - y_1 \end{pmatrix}$$

remains invariant. Let us consider the general case (9). Since for any square matrix it is possible to choose a sufficiently large y_0 such that

$$(x_0 - y_0)^2 - x_2^2 - x_3^2 > 0,$$

there thus exists y_1 such that

$$(x_0 - y_0)^2 = (x_1 - y_1)^2 + x_2^2 + x_3^2.$$

For such y_0, y_1, the relation

$$(-x_0 - y_0)^2 = (x_1 - y_1)^2 + x_2^2 + x_3^2$$

does not hold. For this reason, (a) can hold only for the plus sign. It follows that every element remains invariant.

Up to this point we have made repeated use of affine transformations to reduce Φ to the identity transformation. Therefore Φ is itself an affine transformation. $\square$

5.11 The Fundamental Theorems of Spacetime Geometry

Theorem 1. *A one-to-one correspondence which takes spacetime points onto themselves, and under which the determinant,*

$$|X_1 - X_2|, \tag{1}$$

is invariant, is necessarily a Poincaré transformation, i.e. a transformation generated by:

$$Y = \pm A X \bar{A}' + B, \qquad |A| = 1, \qquad \bar{B}' = B,$$
$$Y = X'.$$

PROOF. A transformation which leaves (1) invariant is, naturally, just a coherence-preserving transformation. Hence

$$Y = \rho A X \bar{A}' + B$$
$$\text{or} \quad Y = \rho A X' \bar{A}' + B.$$

However,

$$|Y_1 - Y_2| = \rho^2 |X_1 - X_2| \cdot |A|^2 = \rho^2 |X_1 - X_2|.$$

We therefore obtain $\rho = \pm 1$. □

Theorem 2. *Under the hypothesis of theorem 1, assume in addition the transformation does not cause time-reversal and preserves spatial orientation; then it is necessarily of the form:*

$$Y = A X \bar{A}' + B.$$

Obviously we also have: the invariance of causal relations is a consequence of the constancy of the speed of light.

5.12 The Projective Geometry of Hermitian Matrices

In the preceding treatment, We have implicitly agreed to "transform H matrices in the finite domain into H matrices in the finite domain". If we allow for an H matrix at infinity, then we shall have the following "projective geometry of H matrices".

First of all, for any H matrix, we introduce "homogeneous coordinates", that is, express H as

$$X = X_2^{-1} X_1, \tag{1}$$

where X_1, X_2 are both 2×2 matrices. The pair of matrices

$$(X_1, X_2) \tag{2}$$

is called the *homogeneous coordinates of X*. Since

$$X = \bar{X}',$$

we have

$$X_2^{-1} X_1 = \bar{X}_1' \bar{X}_2^{-1},$$

i.e.

$$X_1 \bar{X}_2' = X_2 \bar{X}_1', \tag{3}$$

or we may rewrite it as

$$(X_1, X_2)J(\overline{X_1, X_2})' = 0, \qquad J = \begin{pmatrix} 0 & I \\ -I & 0 \end{pmatrix}. \tag{4}$$

Of course we have assumed that X_2 is nonsingular. At the same time, for all Q, $|Q| \neq 0$,

$$Q(X_1, X_2) = (QX_1, QX_2) \tag{5}$$

represents the same X. We shall now expand on this concept. If the rank of the 2×4 matrix (2) is 2, and moreover, if (2) satisfies (4), then (X_1, X_2) is called a *Hermitian pair*, or more simply, an *H pair*. If two H pairs differ by a factor (as in (5)), then they are said to be *equivalent*. H matrices are then partitioned into equivalence classes. The points of the *projective space of H matrices* are by definition these equivalence classes.

Let T be a 4×4 matrix which satisfies

$$TJ\overline{T}' = J. \tag{6}$$

Then transformation

$$(X_1^*, X_2^*) = Q(X_1, X_2)T \tag{7}$$

is called a *transformation of the projective space*. Now, letting

$$T = \begin{pmatrix} A & B \\ C & D \end{pmatrix}, \tag{8}$$

we have

$$A\overline{B}' = B\overline{A}', \qquad C\overline{D}' = D\overline{C}', \qquad A\overline{D}' - B\overline{C}' = I. \tag{9}$$

In nonhomogeneous form, transformation (7) may be rewritten as

$$\begin{aligned} X^* = X_2^{*-1}X_1^* &= (X_1B + X_2D)^{-1}(X_1A + X_2C) \\ &= (XB + D)^{-1}(XA + C) \\ &= (\overline{A}'X + \overline{C}')(\overline{B}'X + \overline{D}')^{-1}; \end{aligned} \tag{10}$$

this can be directly verified from (9). The points satisfying $|X_2| = 0$ are called the *points at infinity*. From (10) we know that

$$\begin{aligned} X^* - Y^* &= (XB + D)^{-1}(XA + C) - (\overline{A}'Y + \overline{C}')(\overline{B}'Y + \overline{D}')^{-1} \\ &= (XB + D)^{-1}[(XA + C)(\overline{B}'Y + \overline{D}') \\ &\quad - (XB + D)(\overline{A}'Y + \overline{C}')](\overline{B}'Y + \overline{D}')^{-1} \\ &= (XB + D)^{-1}(X - Y)(\overline{B}'Y + \overline{D}')^{-1}. \end{aligned} \tag{11}$$

Hence the coherence relation remains invariant (that is if X^*, Y^*, X, Y are all finite points).

In terms of homogeneous coordinates, the coherence relation may be written as

$$|(X_1, X_2)J(\overline{Y_1, Y_2})'| = 0. \tag{12}$$

Let us now prove the following:

Theorem 1. *The set of one-one correspondences of the projective space of H matrices onto itself which preserve the coherence relation* (12) *forms a group, which is generated by the set of all transformations of the form* (7), *where T either satisfies* (6) *or* $TJ\bar{T}' = -J$, *together with the transforms*

$$(X_1^*, X_2^*) = (\bar{X}_1, \bar{X}_2).$$

Since the proof of this theorem is straightforward, it should not be difficult for those readers who are familiar with the fundamental theorem of affine geometry to devise a proof.

Under more stringent assumptions, Fok rather unnecessarily applied the theory of partial differential equations (wave front equations) together with more sophisticated mathematical tools to arrive at the same conclusion. He called this group the *group of Möbius transformations,* that is, in addition to the affine transformations of the Hermitian matrices, we also have those satisfying

$$x_i^* = [x_i - \alpha_i(x_0^2 - x_1^2 - x_2^2 - x_3^2)]/[1 - 2(\alpha_0 x_0 - \alpha_1 x_1 - \alpha_2 x_2 - \alpha_3 x_3)$$
$$+ (\alpha_0^2 - \alpha_1^2 - \alpha_2^2 - \alpha_3^2)(x_0^2 - x_1^2 - x_2^2 - x_3^2)].$$

Applying

$$\begin{pmatrix} \alpha_0 + \alpha_1 & \alpha_2 + i\alpha_3 \\ \alpha_2 - i\alpha_3 & \alpha_0 - \alpha_1 \end{pmatrix}^{-1} = \frac{1}{\alpha_0^2 - \alpha_1^2 - \alpha_2^2 - \alpha_3^2} \begin{pmatrix} \alpha_0 - \alpha_1 & -\alpha_2 - i\alpha_3 \\ -\alpha_2 + i\alpha_3 & \alpha_0 + \alpha_1 \end{pmatrix},$$

it is then not difficult to show that the Möbius transformations may be written in the form of (10).

5.13 Projective Transformations and Causal Relations

At this point it is worth noting the following special case, viz., the *projective transformation* (Möbius transformation):

$$X^* = X\left(\begin{pmatrix} 0 & 0 \\ 0 & -2 \end{pmatrix} X + I\right)^{-1}.$$

This transformation takes 0 into itself and I into $\left(\begin{smallmatrix} 1 & 0 \\ 0 & -1 \end{smallmatrix}\right)$, that is, this transformation destroys the causal relation. Delving even deeper into this case yields the following results:

Any transformation (10) with a nontrivial denominator necessarily destroys the causal relation. Before investigating this further, let us first prove the following lemma:

Lemma. *Let H be a given Hermitian matrix. If for any Hermitian matrix X satisfying* $|X| > 0$ *we have* $|HX + I| > 0$, *then* $H = 0$.

PROOF: Without any loss in generality, we may choose $H = \begin{pmatrix} \alpha & 0 \\ 0 & \beta \end{pmatrix}, (\alpha \geqslant \beta, \ \alpha > 0)$.
Now, if $H \neq 0$, we wish to find an $X = \begin{pmatrix} \xi & 0 \\ 0 & \eta \end{pmatrix}, \ \xi\eta > 0$, such that

$$|XH + I| = (\alpha\xi + 1)(\beta\eta + 1) < 0.$$

Thus,

(1) if $\beta < 0$, then we may choose an ξ to be a sufficiently small positive number, and $\eta > 1/|\beta|$ will do.

(2) if $\beta = 0$, then choose $\xi < -1/\alpha$ and $\eta < 0$ will do.

(3) if $\alpha \geqslant \beta > 0$, then we may choose $\varepsilon > 0$, and

$$\xi = -\frac{1 + \varepsilon}{\alpha}, \qquad \eta = -\frac{1 - \varepsilon}{\beta},$$

such that

$$\xi\eta = \frac{1 - \varepsilon^2}{\alpha\beta} > 0,$$

and

$$(\alpha\xi + 1)(\beta\eta + 1) = [-(1 + \varepsilon) + 1][-(1 - \varepsilon) + 1] = -\varepsilon^2 < 0. \qquad \square$$

Let us now prove:

Theorem 1. *A projective transformation which is not an affine transformation necessarily destroys causal relations.*

PROOF. Assuming that this transformation takes X, Y into X^*, Y^*, then

$$X^* - Y^* = (XB + D)^{-1}(X - Y)(\bar{B}'Y + \bar{D}')^{-1}.$$

Since there is an affine transformation that takes any point into 0, there is no harm in assuming that $Y = Y^* = 0$. Thus $C = 0$ and $|D| \neq 0$. The expression above becomes

$$X^* = (XB + D)^{-1}X\bar{D}'^{-1},$$

and hence

$$|X^*| = |XB + D|^{-1}|\bar{D}'|^{-1}|X|$$
$$= |XB\bar{D}' + I|^{-1}|D\bar{D}'|^{-1}|X|.$$

Since we know from (9) of §12 that BD^{-1} is a Hermitian matrix, we have now proved the theorem using the lemma. $\qquad \square$

5.14 Remarks

(a) A conformal metric. Letting $Y \to X$ in (11) of §12 gives

$$dX^* = (XB + D)^{-1} \, dX(\bar{B}'X + \bar{D}')^{-1}.$$

Taking the determinant, we have

$$dx_0^{*2} - dx_1^{*2} - dx_2^{*2} - dx_3^{*2} = \rho(x_0, x_1, x_2, x_3)^{-2}(dx_0^2 - dx_1^2 - dx_2^2 - dx_3^2),$$

where $\rho = |XB + D|$. This is the origin of the term "conformal".

(b) If we continue to use assumption (A) of §1, but now weaken assumption (B) of §1 to (B') as stated below, we may still reach the same conclusions.

(B') To each observer, the speed of light is constant, i.e. the local version of the constancy of the speed of light.

Let the velocity vector of affine space be (v_1, v_2, v_3). Its rule for transformation is

$$x_i^* = \sum_{j=0}^{3} a_{ij} x_j, \qquad i = 0, 1, 2, 3,$$

and as a result,

$$v_i^* = \frac{dx_i^*}{dx_0^*} = \frac{a_{i0} + \sum_{j=1}^{3} a_{ij} v_j}{a_{00} + \sum_{j=1}^{3} a_{0j} v_j}, \qquad i = 1, 2, 3,$$

i.e. it becomes a 3-dimensional projective space. If the speed of light is constant to each observer, then from

$$v_1^2 + v_2^2 + v_3^2 = 1 \tag{1}$$

(we are letting the speed of light be 1), the invariance of (1) means that any v_1, v_2, v_3 satisfying (1) necessarily satisfies

$$\left(a_{00} + \sum_{j=1}^{3} a_{0j} v_j \right)^2 = \sum_{i=1}^{3} \left(a_{i0} + \sum_{j=1}^{3} a_{ij} v_j \right)^2. \tag{2}$$

Choosing $v_1 = \pm 1$, $v_2 = v_3 = 0$, we obtain

$$(a_{00} \pm a_{01})^2 = \sum_{i=1}^{3} (a_{i0} \pm a_{i1})^2.$$

Successively taking $+$ and $-$ signs, and then adding and subtracting, we obtain

$$a_{00} a_{01} = \sum_{i=1}^{3} a_{i0} a_{i1},$$

$$a_{00}^2 + a_{01}^2 = \sum_{i=1}^{3} (a_{i0}^2 + a_{i1}^2).$$

By the same method we obtain

$$a_{00} a_{0j} = \sum_{i=1}^{3} a_{i0} a_{ij}, \qquad j = 2, 3.$$

$$a_{00}^2 + a_{0j}^2 = \sum_{i=1}^{3} (a_{i0}^2 + a_{ij}^2), \qquad j = 2, 3.$$

We therefore obtain

$$
\begin{cases}
-a_{0j}^2 + \sum_{i=1}^{3} a_{ij}^2 = a_{00}^2 - \sum_{i=1}^{3} a_{i0}^2 = \rho, & j = 1, 2, 3. \\
a_{00}a_{0j} - \sum_{i=1}^{3} a_{i0}a_{ij} = 0, & j = 1, 2, 3,
\end{cases}
$$

that is, we have

$$ X^* = XL, $$

and moreover, $L = L^{(4,4)}$ satisfies

$$ L[1, -1, -1, -1]L' = \rho[1, -1, -1, -1]. $$

(c) All vectors (v_1, v_2, v_3) representing speed less than the speed of light form a space

$$ v_1^2 + v_2^2 + v_3^2 < 1. $$

The transformation group of this space is just the group mentioned in (b). In chapter 7 we shall discuss the 2-dimensional analogue of this space.

CHAPTER 6
Non-Euclidean Geometry

6.1 The Geometric Properties of Extended Space

In Chapter 3 we saw that the group G which we have been discussing is formed from the transformation

$$y = \frac{xT + xx'v_1 + v_2}{xu_2' + xx'b + d} \tag{1}$$

(And at the same time we have

$$yy' = \left(\frac{xu_1' + xx'a + c}{xu_2' + xx'b + d}\right).) \tag{2}$$

Observe that the matrix

$$M = \begin{pmatrix} T & u_1' & u_2' \\ v_1 & a & b \\ v_1 & c & d \end{pmatrix} \tag{3}$$

satisfies

$$MJM' = J. \tag{4}$$

Thus in terms of homogeneous coordinates we have

$$(\xi^*, \eta_1^*, \eta_2^*) = \rho(\xi, \eta_1, \eta_2)M, \tag{5}$$

where M is a transformation leaving invariant

$$\xi\xi' - \eta_1\eta_2 = 0.$$

Letting $\eta_1 = s_1 + s_2, \eta_2 = -s_1 + s_2$ then gives

$$\xi\xi' + s_1^2 - s_2^2 = 0,$$

and dividing this by s_2 we obtain an $(n + 1)$-dimensional unit sphere. There-fore the study of the n-dimensional space expanded through the group G is equivalent to the study of the spherical geometry of the unit sphere in $(n + 1)$-dimensional space. We shall discuss this type of geometry again when we study mixed partial differential equations later on. However, it may be mentioned that this is just a generalization of the method of stereographic projection which produces a correspondence between the complex plane and the unit sphere.

We also know that under the transformation of G, the spheres are divided into three types: real spheres, point spheres and imaginary spheres. More-over, these three types may all be transformed into the following canonical forms:

(i) $xx' = 1$ (real sphere)
(ii) $xx' = 0$ (point sphere)
(iii) $xx' = -1$ (imaginary sphere)

Fix a sphere to be the *absolute sphere*, then those transformations under which this sphere remains invariant form a group which we denote by H. The geometry within the group H is called non-Euclidean geometry, and in correspondence to (i), (ii) and (iii), may be divided into three geometric cate-gories: hyperbolic, parabolic and elliptic.

In order to conveniently see the applications of expression (4), we first write out the relations included therein:

$$TT' - \tfrac{1}{2}(u'_1 u_2 + u'_2 u_1) = I^{(n)}, \tag{6}$$
$$Tv'_1 - \tfrac{1}{2}(u'_1 b + u'_2 a) = 0, \tag{7}$$
$$Tv'_2 - \tfrac{1}{2}(u'_1 d + u'_2 c) = 0, \tag{8}$$
$$v_1 v'_1 - ab = 0, \tag{9}$$
$$v_1 v'_2 - \tfrac{1}{2}(ad + bc) = -\tfrac{1}{2}, \tag{10}$$
$$v_2 v'_2 - cd = 0. \tag{11}$$

Taking the inverse of (4) we have

$$M'J^{-1}M = J^{-1}, \qquad J^{-1} = \begin{pmatrix} I & 0 & 0 \\ 0 & 0 & -2 \\ 0 & -2 & 0 \end{pmatrix}.$$

That is to say, we obtain

$$T'T - 2(v'_1 v_2 + v'_2 v_1) = I, \tag{12}$$
$$T'u'_1 - 2(v'_1 c + v'_2 a) = 0, \tag{13}$$
$$T'u'_2 - 2(v'_1 d + v'_2 b) = 0, \tag{14}$$
$$u_1 u'_1 - 4ac = 0, \tag{15}$$
$$u_1 u'_2 - 2(ad + bc) = -2, \tag{16}$$
$$u_2 u'_2 - 4bd = 0. \tag{17}$$

6.2 Parabolic Geometry

Parabolic geometry is the geometry defined by the group under the action of which a point space remains invariant. Since we may assume that this point is $x = \infty$, we know from (1.1) then, that $b = 0$; and again from (1.9) and (1.17) we know that

$$v_1 = 0, \qquad u_2 = 0. \tag{1}$$

Again from (1.6),

$$TT' = I, \tag{2}$$

and since from (1.6) $ad = 1$, then

$$a = 1/d.$$

Once again from (1.8),

$$u_1 = (2/d)v_2 T', \tag{3}$$

and finally from (1.15),

$$c = (4/d)v_2 v_2'. \tag{4}$$

(1), (2), (3) and (4) completely determine the form of the transformation, that is

$$M = \begin{pmatrix} T & \dfrac{2}{d}Tv_2' & 0 \\ 0 & \dfrac{1}{d} & 0 \\ v_2 & \dfrac{4}{d}v_2 v_2' & d \end{pmatrix},$$

in other words, we have obtained the general form of the transformation in parabolic geometry,

$$y = a(xT + v). \tag{5}$$

This is the composite of rotation (and inversion)

$$y = xT,$$

and translation

$$y = x + v,$$

and magnification (and contraction)

$$y = ax.$$

Parabolic geometry is precisely defined by the following entities:
space: all the finite points.
group: all groups which are generated by rotation, inversion, translation, magnification and contraction.

If we disregard magnification and contraction, then this geometry is the n-dimensional Euclidean geometry. We shall not discuss this type of geometry in depth.

6.3 Elliptical Geometry

The imaginary sphere may be represented by the vector

$$(0; 1, 1).$$

Our group is generated by the square matrices M which satisfy

$$(0; 1, 1) = \rho(0; 1, 1)M'.$$

Making the change of variables

$$\eta_1 = s_1 + s_2, \qquad \eta_2 = -s_1 + s_2,$$

we change M into N so that we have

$$N[1, 1, \ldots, 1, -1]N' = [1, \ldots, 1, -1]; \tag{1}$$

moreover the vector $(0; 1, 1)$ becomes

$$(0; 0, 1)$$

$(s_1 = 0, s_2 = 1)$, that is,

$$(0, 0, \ldots, 0, 1)N = \rho(0, 0, \ldots, 0, 1).$$

Therefore

$$N = \begin{pmatrix} N_1^{(n+1)} & w'_1 \\ 0 & d \end{pmatrix},$$

and from (1) we obtain $w_1 = 0, d = \pm 1$ and

$$N_1 N'_1 = I^{(n+1)},$$

where N_1 is an $(n + 1) \times (n + 1)$ orthogonal matrix.

Thus, elliptical geometry is the geometry of the $(n + 1)$-dimensional sphere and it is the geometry of the group T generated by rotation and inversion.

6.4 Hyperbolic Geometry

The vector

$$(0, \ldots, 0; 1, -1)$$

represents the unit ball. Our group is generated by the square matrices M which satisfy

$$(0, \ldots, 0, 1, -1) = \rho(0, 0, \ldots, 0, 1, -1)M',$$

that is to say, it is generated by the transformations which satisfy

$$u_1 = u_2 \tag{1}$$

and
$$a - b + c - d = 0. \tag{2}$$

This can also be deduced from (1.2) by taking $xx' = yy' = 1$.

By letting $u_1 = u_2 = u$, we know from (1.6) that

$$TT' = I^{(n)} + u'u, \tag{3}$$

that is T is nonsingular. Again from (1.7) and (1.8) we have

$$\begin{cases} v_1 = \frac{1}{2}(a + b)uT'^{-1} \\ v_2 = \frac{1}{2}(c + d)uT'^{-1} \end{cases} \tag{4}$$

And again from (1.9) and (3) we know that

$$\tfrac{1}{4}(a + b)^2 uT'^{-1}T^{-1}u = ab,$$

i.e.

$$(a + b)^2 u(I + u'u)^{-1}u' = 4ab$$
$$(a + b)^2 uu'(1 + uu')^{-1} = 4ab,$$

i.e. we get

$$(a - b)^2 uu' = 4ab. \tag{5}$$

By similar methods, from (1.10) and (1.11) we obtain

$$[(a - b)(c - d) + 2]uu' = -2 + 2(ad + bc) \tag{6}$$
$$(c - d)^2 uu' = 4cd \tag{7}$$

From (2), (5), (6) and (7) we obtain the solutions

$$\begin{cases} a = d = \frac{1}{2}(\pm 1 \pm \sqrt{1 + uu'}), \\ b = c = \frac{1}{2}(\mp 1 \pm \sqrt{1 + uu'}). \end{cases} \tag{8}$$

Since the case for $u = 0$ is very easily dealt with, we will assume that $u \neq 0$. Furthermore, since $\pm M$ represent the same transformation, we may also assume that $a > 0$, that is,

$$\begin{cases} a = d = \frac{1}{2}(\pm 1 + \sqrt{1 + uu'}), \\ b = c = \frac{1}{2}(\mp 1 + \sqrt{1 + uu'}). \end{cases}$$

Again from

$$1 - yy' = \frac{(a - b)(1 - xx')}{xu' + xx'b + a},$$

we know that the transformation which takes the unit ball into itself is

$$\begin{aligned} a = d = \tfrac{1}{2}(1 + \sqrt{1 + uu'}), \\ b = c = \tfrac{1}{2}(-1 + \sqrt{1 + uu'}), \end{aligned} \tag{9}$$

that is to say, any entry of M must satisfy (3), and v_1, v_2, a, b, c and d are explicitly given by (4) and (9).

This not only computes that in this group there are n such parameters u, but also, that if u is fixed, T has $\frac{1}{2}n(n - 1)$ parameters. That is, there are altogether $\frac{1}{2}n(n + 1)$ parameters. (The number of parameters of orthogonal matrices is $\frac{1}{2}n(n - 1)$.)

In summing up, the transformation within the group G which takes the unit ball into itself may be written as

$$y = \frac{xT + \frac{1}{2}(1 + xx')\sqrt{1 + uu'}\,uT'^{-1}}{xu' + \frac{1}{2}(1 + xx')\sqrt{1 + uu'} + \frac{1}{2}(1 - xx')}. \tag{10}$$

However, since

$$uT'^{-1} = u(I + u'u)^{-1}T = (1 + uu')^{-1}uT,$$

(10) may therefore be rewritten as

$$y = \frac{x + \frac{1}{2}(1 + xx')(1 + uu')^{-1/2}u}{xu' + \frac{1}{2}(1 + xx')(1 + uu')^{1/2} + \frac{1}{2}(1 - xx')}\,T.$$

The transformation which leaves the point 0 invariant is the transformation $u = 0$, that is,

$$y = xT, \qquad TT' = I. \tag{11}$$

Hence the general transformation is indeed generated by the transformations discussed in Chapter 1. Since the transformations in Chapter 1 can take an arbitrary point a into 0, it is sufficient to consider only the general form of those transformations which take 0 into 0; and this general form is just (10), which was mentioned in Chapter 1.

6.5 Geodesics

Theorem 1. *Given any two points, there is one and only one geodesic passing through them. Precisely, let x_0, x_0^* be two points, then the integral*

$$\int_{x_0}^{x_0^*} \frac{\sqrt{dx\,dx'}}{1 - xx'}$$

taken along the geodesic is smaller than that taken along any other curve joining x_0 and x_0^*.

PROOF. Without loss in generality, we may choose $x_0 = 0$ and

$$x_0^* = (\delta, 0, \ldots, 0, 0),$$

for from transitivity we may assume $x_0 = 0$, and by making one rotation we may then further assume that $x_0^* = (\delta, 0, \ldots, 0)$.

Let

$$x = x(t), \qquad 0 \leqslant t \leqslant 1,$$

be a curve connecting these two points, i.e.

$$x(0) = 0, \qquad x(1) = (\delta, 0, \ldots, 0).$$

The integral is then equal to

$$\int_0^1 \frac{\sqrt{\left(\dfrac{dx_1}{dt}\right)^2 + \cdots + \left(\dfrac{dx_n}{dt}\right)^2}}{1 - x_1^2 \cdots - x_n^2}\, dt \geqslant \int_0^1 \frac{\dfrac{dx_1}{dt}}{1 - x_1^2}\, dt$$

$$= \frac{1}{2} \log \frac{1 + x_1(t)}{1 - x_1(t)}\bigg|_0^1 = \frac{1}{2} \log \frac{1 + \delta}{1 - \delta},$$

which is what we wished to prove. $\qquad\qquad\qquad\qquad\square$

CHAPTER 7
Partial Differential Equations of Mixed Type

We will begin with the 2-dimensional case, but those readers familiar with linear algebra can easily generalize to the higher dimensional cases.

7.1 Real Projective Planes

Always staying with the unit circle, let us first study the projective transformations which leave the unit circle invariant.

The group of transformations we wish to discuss is formed by the projective transformations which take

$$x^2 + y^2 < 1 \tag{1}$$

into itself, i.e. the transformations

$$x_1 = \frac{a_1 x + b_1 y + c_1}{a_3 x + b_3 y + c_3}, \qquad y_1 = \frac{a_2 x + b_2 y + c_2}{a_3 x + b_3 y + c_3}, \tag{2}$$

such that the matrix

$$A = \begin{pmatrix} a_1 & b_1 & c_1 \\ a_2 & b_2 & c_2 \\ a_3 & b_3 & c_3 \end{pmatrix}$$

is nonsingular, and

$$A' \begin{pmatrix} 1 & 0 & 0 \\ 0 & 1 & 0 \\ 0 & 0 & -1 \end{pmatrix} A = \rho \begin{pmatrix} 1 & 0 & 0 \\ 0 & 1 & 0 \\ 0 & 1 & -1 \end{pmatrix}. \tag{3}$$

Due to the homogeneity of (2), one may assume that $\rho = \pm 1$; then taking the determinant of (3), one sees that $\rho = 1$. So from now on we shall assume that $\rho = 1$.

We assert that this group, which we will call Γ, is generated by the following elements:

$$\begin{pmatrix} \cos\theta & \sin\theta & 0 \\ -\sin\theta & \cos\theta & 0 \\ 0 & 0 & 1 \end{pmatrix}, \begin{pmatrix} 1 & 0 & 0 \\ 0 & -1 & 0 \\ 0 & 0 & 1 \end{pmatrix}, \begin{pmatrix} 1 & 0 & 0 \\ 0 & \cosh\psi & \sinh\psi \\ 0 & \sinh\psi & \cosh\psi \end{pmatrix};$$

these elements are called rotation, reflection and hyperbolic rotation. Or, somewhat more concretely, these elements can be rewritten as

$$\begin{cases} x_1 = x\cos\theta + y\sin\theta, \\ y_1 = -x\sin\theta + y\cos\theta, \end{cases} \tag{4}$$

$$\begin{cases} x_1 = x, \\ y_1 = -y. \end{cases} \tag{5}$$

and for a real number $\mu\ (-1 < \mu < 1)$,

$$\begin{cases} x_1 = \sqrt{1-\mu^2}\,x/(1-\mu y), \\ y_1 = (y-\mu)/(1-\mu y), \end{cases} \tag{6}$$

or

$$x_1 = x/(y\sinh\psi + \cosh\psi)$$
$$y_1 = (y\cosh\psi + \sinh\psi)/(y\sinh\psi + \cosh\psi).$$

This assertion can be proved as follows. Multiply the right and left sides of A by a square matrix of the form (4), thereby causing $b_1 = a_2 = 0$. In this way it is not difficult to deduce that A becomes

$$\begin{pmatrix} 1 & 0 & 0 \\ 0 & b_2 & c_2 \\ 0 & b_3 & c_3 \end{pmatrix},$$

and this, clearly, is just a product of matrices of types (5) and (6).

Under the action of the group Γ, a point in the disc can be transformed into any point in the disc. The proof of this fact is not difficult; first of all, by rotation any point can be transformed into $(0, \lambda)$ $(\lambda > 0)$. Since $\lambda < 1$, this point can be transformed into $(0, 0)$ by (6). (In fact, the points outside the disc also form a set transitive under Γ.)

We shall now study the differential invariants under the group Γ. The straight line connecting (x, y) and $(x + dx, y + dy)$ is

$$(x + \lambda\,dx, y + \lambda\,dy),$$

and the points of intersection between this straight line and the unit circle may be determined from

$$(x + \lambda\,dx)^2 + (y + \lambda\,dy)^2 = 1,$$

that is,

$$\lambda^2(dx^2 + dy^2) + 2\lambda(x\,dx + y\,dy) + x^2 + y^2 - 1 = 0.$$

The discriminant of this expression is

$$(x\,dx + y\,dy)^2 - (dx^2 + dy^2)(x^2 + y^2 - 1),$$

that is,

$$(1 - y^2)\,dx^2 + 2xy\,dx\,dy + (1 - x^2)\,dy^2.$$

The preceding suggests that this second degree form is probably covariant; and by actual calculation, this is indeed covariant. Moreover, the expression

$$\frac{(1 - y^2)\,dx^2 + 2xy\,dx\,dy + (1 - x^2)\,dy^2}{(1 - x^2 - y^2)^2} \tag{A}$$

is invariant under Γ. This fact can of course be derived from the consideration of the cross ratio (i.e. from the cross ratio of these two points with the two points of intersection on the circle), however, it can also be proved directly as follows:

The numerator of (A) is equal to

$$dx^2 + dy^2 - (y\,dx - x\,dy)^2,$$

and this is clearly invariant under rotation and reflection. Furthermore, it is covariant under (6) because

$$dx_1 = \frac{\sqrt{1 - \mu^2}}{1 - \mu y}\,dx + \frac{\mu\sqrt{1 - \mu^2}}{(1 - \mu y)^2}\,x\,dy,$$

$$dy_1 = \frac{1 - \mu^2}{(1 - \mu y)^2}\,dy,$$

so that

$$dx_1^2 + dy_1^2 - (x_1\,dy_1 - y_1\,dx_1)^2$$

$$= \frac{1}{(1 - \mu y)^4}\left[(1 - \mu^2)\left\{(1 - \mu y)\,dx + \mu x\,dy\right\}^2 + (1 - \mu^2)^2\,dy^2\right]$$

$$- \frac{1 - \mu^2}{(1 - \mu y)^6}\left[x(1 - \mu^2)\,dy - (y - \mu)\left\{(1 - \mu y)\,dx + \mu x\,dy\right\}\right]^2$$

$$= \frac{1 - \mu^2}{(1 - \mu y)^4}\left[\left\{(1 - \mu y)\,dx + \mu x\,dy\right\}^2 + (1 - \mu^2)\,dy^2 - \left\{x\,dy - (y - \mu)\,dx\right\}^2\right]$$

$$= \frac{(1 - \mu^2)^2}{(1 - \mu y)^4}\left[(1 - y^2)\,dx^2 + 2xy\,dx\,dy + (1 - x^2)\,dy^2\right]. \tag{7}$$

On the other hand,

$$1 - x_1^2 - y_1^2 = 1 - \frac{1}{(1 - \mu y)^2}\left[(1 - \mu^2)x^2 + (y - \mu)^2\right]$$

$$= \frac{1 - \mu^2}{(1 - \mu y)^2}\,(1 - x^2 - y^2), \tag{8}$$

hence we obtain

$$\frac{(1 - y_1^2)\, dx_1^2 + 2x_1 y_1\, dx_1\, dy_1 + (1 - x_1^2)\, dx_1^2}{(1 - x_1^2 - y_1^2)^2}$$

$$= \frac{(1 - y^2)\, dx^2 + 2xy\, dx\, dy + (1 - x^2)\, dx^2}{(1 - x^2 - y^2)^2}. \qquad (9)$$

We take this invariant differential form to be our *Riemannian metric*.

7.2 Partial Differential Equations

The following second order partial differential operator is "dual" to (A):

$$\Delta u = (1 - x^2 - y^2)\Bigg[(1 - x^2)\frac{\partial^2}{\partial x^2} - 2xy\frac{\partial^2}{\partial x\, \partial y}$$

$$+ (1 - y^2)\frac{\partial^2}{\partial y^2} - 2x\frac{\partial}{\partial x} - 2y\frac{\partial}{\partial y}\Bigg]u. \qquad (B)$$

We claim this is also invariant under Γ.

There are two ways of proving this. One is by looking at (A) as a Riemannian metric. The *Lamé operator* (or *Beltrami operator*) of this *Riemannian space* is then just (B) and the above property then follows from general considerations. However, this calculation is long and tedious, and by contrast, the method of direct substitution is much easier. We will therefore use the latter method. Since it is easily proved that this operator is invariant under transformations (1.4) and (1.5), we will now prove that it is also invariant under (1.6). We have

$$\frac{\partial u}{\partial x} = \frac{\partial u}{\partial x_1}\frac{\sqrt{1 - \mu^2}}{1 - \mu y},$$

$$\frac{\partial u}{\partial y} = \frac{\partial u}{\partial x_1}\frac{\mu x\sqrt{1 - \mu^2}}{(1 - \mu y)^2} + \frac{\partial u}{\partial y_1}\frac{1 - \mu^2}{(1 - \mu y)^2}.$$

Therefore

$$\frac{\partial^2 u}{\partial x^2} = \frac{\partial^2 u}{\partial x_1^2}\frac{1 - \mu^2}{(1 - \mu y)^2},$$

$$\frac{\partial^2 u}{\partial x\, \partial y} = \frac{\partial^2 u}{\partial x_1^2}\frac{\mu(1 - \mu^2)x}{(1 - \mu y)^3} + \frac{\partial^2 u}{\partial x_1\, \partial y_1}\frac{(1 - \mu^2)^{3/2}}{(1 - \mu y)^3}$$

$$+ \frac{\partial u}{\partial x_1}\frac{\mu\sqrt{1 - \mu^2}}{(1 - \mu y)^2},$$

and

$$\frac{\partial^2 u}{\partial y^2} = \frac{\partial^2 u}{\partial x_1^2}\frac{\mu^2(1-\mu^2)x^2}{(1-\mu y)^4} + 2\frac{\partial^2 u}{\partial x_1\,\partial y_1}\frac{\mu x(1-\mu^2)^{3/2}}{(1-\mu y)^4}$$

$$+ \frac{\partial^2 u}{\partial y_1^2}\frac{(1-\mu^2)^2}{(1-\mu y)^4} + 2\frac{\partial u}{\partial x_1}\frac{\mu^2 x\sqrt{1-\mu^2}}{(1-\mu y)^3} + 2\frac{\partial u}{\partial y_1}\frac{\mu(1-\mu^2)}{(1-\mu y)^3}.$$

Thus

$$(1-x^2)\frac{\partial^2 u}{\partial x^2} - 2xy\frac{\partial^2 u}{\partial x\,\partial y} + (1-y^2)\frac{\partial^2 u}{\partial y^2} - 2x\frac{\partial u}{\partial x} - 2y\frac{\partial u}{\partial y}$$

$$= \frac{\partial^2 u}{\partial x_1^2}\left[(1-x^2)\frac{1-\mu^2}{(1-\mu y)^2} - \frac{2\mu(1-\mu^2)x^2 y}{(1-\mu y)^3} + (1-y^2)\frac{\mu^2 x^2(1-\mu^2)}{(1-\mu y)^4}\right]$$

$$+ 2\frac{\partial^2 u}{\partial x_1\partial y_1}\left[-\frac{(1-\mu^2)^{3/2}xy}{(1-\mu y)^3} + \frac{\mu(1-\mu^2)^{3/2}}{(1-\mu y)^4}(1-y^2)\right]$$

$$+ \frac{\partial^2 u}{\partial y_1^2}(1-y)^2\frac{(1-\mu^2)^2}{(1-\mu y)^4} - 2\frac{\partial u}{\partial x_1}\left[\frac{xy\mu\sqrt{1-\mu^2}}{(1-\mu y)^2}\right.$$

$$\left. - \frac{\mu^2 x^2\sqrt{1-\mu^2}}{(1-\mu y)^3}(1-y^2) + x\frac{\sqrt{1-\mu^2}}{1-\mu y} + y\frac{\mu x\sqrt{1-\mu^2}}{(1-\mu y)^2}\right]$$

$$- 2\frac{\partial u}{\partial y^1}\left[-\frac{(1-\mu^2)\mu}{(1-\mu y)^3}(1-y^2) + \frac{(1-\mu^2)y}{(1-\mu y)^2}\right]$$

$$= \frac{1-\mu^2}{(1-\mu y)^2}\left\{\frac{\partial^2 u}{\partial x_1^2}\left[1 - x^2 - \frac{2\mu yx^2}{1-\mu y} + (1-y^2)\frac{\mu^2 x^2}{(1-\mu y)^2}\right]\right.$$

$$- 2\frac{\partial^2 u}{\partial x_1\,\partial y_1}\left[\frac{(1-\mu^2)^{1/2}x}{1-\mu y}\left(y - \frac{\mu(1-y^2)}{1-\mu y}\right)\right] + \frac{\partial^2 u}{\partial y_1^2}\frac{(1-y^2)(1-\mu^2)}{(1-\mu y)^2}$$

$$- 2\frac{\partial u}{\partial x_1}(1-\mu^2)^{-1/2}x\left[(1-\mu y) + 2\mu y - \frac{\mu^2(1-y^2)}{1-\mu y}\right]$$

$$\left. - 2\frac{\partial u}{\partial y_1}\left[y - \frac{\mu}{1-\mu y}(1-y^2)\right]\right\}$$

$$= \frac{1-\mu^2}{(1-\mu y)^2}\left\{\frac{\partial^2 u}{\partial x_1^2}\left(1 - \frac{(1-\mu^2)x^2}{(1-\mu y)^2}\right) - 2\frac{\partial^2 u}{\partial x_1\,\partial y_1}\frac{(1-\mu^2)^{1/2}x(y-\mu)}{(1-\mu y)^2}\right.$$

$$\left. + \frac{\partial^2 u}{\partial y_1^2}\left[1 - \left(\frac{y-\mu}{1-\mu y}\right)^2\right] - 2\frac{\partial u}{\partial x_1}\frac{(1-\mu^2)^{1/2}x}{1-\mu y} - 2\frac{\partial u}{\partial y_1}\frac{y-\mu}{1-\mu y}\right\}$$

$$= \frac{1-\mu^2}{(1-\mu y)^2}\left\{(1-x_1^2)\frac{\partial^2 u}{\partial x_1^2} - 2x_1 y_1\frac{\partial^2 u}{\partial x_1\,\partial y_1}\right.$$

$$\left. + (1-y_1^2)\frac{\partial^2 u}{\partial y_1^2} - 2x_1\frac{\partial u}{\partial x_1} - 2y\frac{\partial u}{\partial y_1}\right\}.$$

From (1.8) we obtain the invariance property of (B).

Note that the Jacobian of this transformation is

$$\frac{\partial(x_1, y_1)}{\partial(x, y)} = \left(\frac{\sqrt{1 - \mu^2}}{1 - \mu y}\right)^3.$$

If we consider the expressions (A) and (B) *without* the factor $(1 - x^2 - y^2)$, then we should observe that their covariant factor is a non-integral power of the preceding Jacobian, while the partial differential equation

$$(1 - x^2)\frac{\partial^2 u}{\partial x^2} - 2xy\frac{\partial^2 u}{\partial x \partial y} + (1 - y^2)\frac{\partial^2 u}{\partial y^2} - 2\left(x\frac{\partial u}{\partial x} + y\frac{\partial u}{\partial y}\right) = 0 \quad \text{(C)}$$

is an equation with (A) as characteristic lines, which are tangents to the unit circle.

This equation is of mixed type, for in the unit disc it is of elliptic type and outside the unit circle it is of hyperbolic type. The unit circle is the *type-changing curve* (in a self-explanatory sense).

Since under a projective transformation the unit circle is equivalent to any real nonsingular quadratic curve, we are in fact dealing with a problem where a partial differential equation changes type on a real nonsingular quadratic curve.

The polar coordinate form of equation (C) is

$$(1 - \rho^2)\frac{\partial^2 u}{\partial \rho^2} + \frac{1}{\rho^2}\frac{\partial^2 u}{\partial \theta^2} + \left(\frac{1}{\rho} - 2\rho\right)\frac{\partial u}{\partial \rho} = 0, \quad \text{(D)}$$

which may also be written as

$$\rho|1 - \rho^2|^{1/2}\frac{\partial}{\partial \rho}\left(\frac{\rho(1 - \rho^2)}{|1 - \rho^2|^{1/2}}\frac{\partial u}{\partial \rho}\right) + \frac{\partial^2 u}{\partial \theta^2} = 0.$$

Let $\mathcal{D}$ be a domain on the projective plane, if a function $u = u(x, y)$ on $\mathcal{D}$ satisfies partial differential equations (C) or (D), then $u(x, y)$ may be called a harmonic function on $\mathcal{D}$; the additional conditions which $u(x, y)$ must satisfy in order to qualify for this appellation will be discussed later.

Since the Lamé operator originates from the consideration of potential functions in curvilinear coordinates, the introduction of this concept in this manner is very natural.

If $\mathcal{D}$ lies in the unit disc, then we get the usual elliptic equation. If $\mathcal{D}$ lies completely outside the unit circle, then we just get the usual hyperbolic partial differential equation. We will now focus our attention on the case where part of $\mathcal{D}$ is in the unit disc and part of it is outside the unit disc. In other words, we will now discuss partial differential equations of mixed type.

7.3 Characteristic Curves

The solutions of the differential equation

$$(1 - y^2)\,dx^2 + 2xy\,dx\,dy + (1 - x^2)\,dy^2 = 0 \qquad (1)$$

are called *characteristic lines.* Let us now find the solutions to this equation.

$x = 1$, which is a tangent to the unit circle, is obviously one solution to (1). Now since it is possible to obtain all tangents to the unit circle under the action of the group Γ, we know that these tangents all satisfy differential equation (1); that is to say they comprise the general solution to (1), whereas the unit circle is the singular solution to this equation.

The tangents of the unit circle are characteristic lines, and the envelope of these characteristic lines is just the unit circle.

The general form of these characteristic lines is

$$x \cos \alpha + y \sin \alpha = 1,$$

so that

$$y^2(1 - \cos^2 \alpha) = (1 - x \cos \alpha)^2,$$
$$(x^2 + y^2) \cos^2 \alpha - 2x \cos \alpha + 1 - y^2 = 0,$$

and therefore,

$$\cos \alpha = \frac{x \pm \sqrt{x^2 - (x^2 + y^2)(1 - y^2)}}{x^2 + y^2}$$

$$= \frac{x \pm y\sqrt{x^2 + y^2 - 1}}{x^2 + y^2}.$$

Thus we have derived that equation (2) has a general solution of the form

$$u(x, y) = f_1\left(\frac{x + y\sqrt{x^2 + y^2 - 1}}{x^2 + y^2}\right) + f_2\left(\frac{x - y\sqrt{x^2 + y^2 - 1}}{x^2 + y^2}\right),$$

where f_1, f_2 are two arbitrary functions. In polar coordinates the general solution becomes

$$g_1\left(\theta + \cos^{-1}\frac{1}{\rho}\right) + g_2\left(\theta - \cos^{-1}\frac{1}{\rho}\right).$$

If

$$u(x, y)$$

is a solution to equation (C), then so is

$$u(x \cos \psi + y \sin \psi, -x \sin \psi + y \cos \psi),$$

and furthermore, so is

$$u_1(x, y) = \frac{1}{2\pi} \int_0^{2\pi} u(x \cos \psi + y \sin \psi, -x \sin \psi + y \cos \psi)\,d\psi.$$

This type of solution is clearly invariant under rotation and this function is obviously a function of ρ alone and is independent of θ. Let us first consider this type of solution; from (D) we have:

$$\frac{\partial}{\partial \rho}\left(\frac{\rho(1-\rho^2)}{|1-\rho^2|^{1/2}}\frac{\partial u}{\partial \rho}\right) = 0, \qquad \frac{\rho(1-\rho^2)}{|1-\rho^2|^{1/2}}\frac{\partial u}{\partial p} = C_1,$$

$$u = \begin{cases} C_1 \log \dfrac{1+\sqrt{1-\rho^2}}{\rho} + C_2, & \text{for } \rho \leqslant 1, \\[3mm] C_1 \arccos \dfrac{1}{\rho} + C_2, & \text{for } \rho > 1. \end{cases}$$

7.4 The Relationship Between this Partial Differential Equation and Lav'rentiev's Equation

By making the change of variables $\xi = f(\rho)$ in the polar coordinates equation

$$\rho^2(1-\rho^2)\frac{\partial^2 u}{\partial \rho^2} + \rho(1-2\rho^2)\frac{\partial u}{\partial \rho} + \frac{\partial^2 u}{\partial \theta^2} = 0, \qquad \text{(D)}$$

we obtain

$$\frac{\partial u}{\partial \rho} = \frac{\partial u}{\partial \xi}f'(\rho), \qquad \frac{\partial^2 u}{\partial \rho^2} = \frac{\partial^2 u}{\partial \xi^2}f'^2(\rho) + \frac{\partial u}{\partial \xi}f''(\rho),$$

so that substitution into (D) yields

$$\rho^2(1-\rho^2)f'^2(\rho)\frac{\partial^2 u}{\partial \xi^2} + [\rho^2(1-\rho^2)f''(\rho) + \rho(1-2\rho^2)f'(\rho)]\frac{\partial u}{\partial \xi} + \frac{\partial^2 u}{\partial \theta^2} = 0.$$

We now choose $f(\rho)$ such that

$$\rho^2|1-\rho^2|(f'(\rho))^2 = 1. \qquad \text{(1)}$$

Differentiating this expression gives us

$$2f'(\rho)f''(\rho)\rho^2|1-\rho^2| + \{2\rho|1-\rho^2| - 2\rho^2 \operatorname{sgn}(1-\rho^2)\rho\}f'^2(\rho) = 0,$$

that is,

$$\rho^2(1-\rho^2)f''(\rho) + \rho(1-2\rho^2)f'(\rho) = 0. \qquad \text{(2)}$$

If $f(\rho)$ satisfies (1), then expression (D) becomes

$$\rho^2(1-\rho^2)f'^2(\rho)\frac{\partial^2 u}{\partial \xi^2} + \frac{\partial^2 u}{\partial \theta^2} = 0,$$

that is,

$$\operatorname{sgn}(1-\rho^2)\frac{\partial^2 u}{\partial \xi^2} + \frac{\partial^2 u}{\partial \theta^2} = 0. \qquad \text{(3)}$$

We will now solve expression (1); choose $\xi = f(\rho)$ such that

$$\frac{d\xi}{d\rho} = -\frac{1}{\rho|1 - \rho^2|^{1/2}};$$

the solution of this then is

$$\xi = \begin{cases} \log \dfrac{1 + \sqrt{1 - \rho^2}}{\rho} + C, & \text{for } \rho < 1, \\[2mm] -\arccos \dfrac{1}{\rho} + C', & \text{for } \rho > 1. \end{cases}$$

Choose $C = C' = 0$, so that we obtain the transformation

$$\xi = \begin{cases} \cosh^{-1} \dfrac{1}{\rho}, & \text{for } 0 \leqslant \rho \leqslant 1, \\[2mm] -\cos^{-1} \dfrac{1}{\rho}, & \text{for } \rho \geqslant 1. \end{cases} \qquad \text{(E)}$$

This transformation is continuous, and its general behavior is:

ρ	0	1	∞
ξ	∞	0	$-\dfrac{\pi}{2}$

Under this transformation, the partial differential equation under consideration then becomes

$$\operatorname{sgn} \xi \, \frac{\partial^2 u}{\partial \xi^2} + \frac{\partial^2 u}{\partial \theta^2} = 0,$$

where $-\pi/2 \leqslant \xi \leqslant \infty$ and $-\pi < \theta \leqslant \pi$.

From (E) we have

$$\rho = \begin{cases} \dfrac{1}{\cosh \xi}, & \text{for } \xi \geqslant 0, \\[2mm] \dfrac{1}{\cos \xi}, & \text{for } -\dfrac{\pi}{2} \leqslant \xi \leqslant 0. \end{cases} \qquad \text{(E')}$$

The transformation is continuous and moreover has continuous first order derivatives. However, its second order derivatives are not continuous at $\xi = 0$.

This clearly shows that when studying the Lav'rentiev equation, we cannot assume the existence of continuous second order derivatives.

If we consider (ρ, θ) and $(\xi, 0)$ as rectangular coordinates, then we have the following figures:

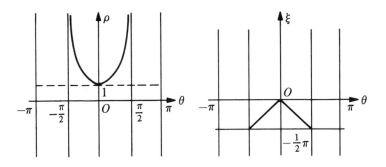

The region under consideration in the (ρ, θ) plane is the half-strip $\rho > 0$, $-\pi < \theta \leqslant \pi$; the part which satisfies $0 < \rho \leqslant 1$ is an elliptical region, the part which satisfies $\rho > 1$ is a hyperbolic region, and the U-shaped curve is a characteristic curve. The other characteristic curves are obtained by horizontal translation of this curve. However it must be noted that if we identify the two straight lines $\theta = \pi$ and $\theta = -\pi$ in the (ξ, θ) plane, then the regions under consideration are $-\frac{1}{2}\pi < \xi$ and $-\pi < \theta \leqslant \pi$, where the part $\xi < 0$ is a hyperbolic region and the $\wedge$-shaped curve is one of the characteristic curves. By horizontal translation we can again obtain the other characteristic curves.

7.5 Separation of Variables

Let us now study the equation

$$\rho^2(1 - \rho^2)\frac{\partial^2 u}{\partial \rho^2} + \rho(1 - 2\rho^2)\frac{\partial u}{\partial \rho} = -\frac{\partial^2 u}{\partial \theta^2}. \tag{D}$$

If the solution is of the form $u = \varphi(\rho)\psi(\theta)$, then

$$\frac{\rho^2(1 - \rho^2)\varphi''(\rho) + \rho(1 - 2\rho^2)\varphi'(\rho)}{\varphi(\rho)} = -\frac{\psi''(\theta)}{\psi(\theta)}. \tag{1}$$

From the periodicity of θ we know that

$$-\frac{\psi''(\theta)}{\psi(\theta)} = n^2, \tag{2}$$

where n is an integer, and whence either $\psi(\theta) = \cos n\theta$ or $\psi(\theta) = \sin n\theta$. Combining (1) and (2) gives

$$\rho^2(1 - \rho^2)\varphi'' + \rho(1 - 2\rho^2)\varphi' - n^2\varphi = 0. \tag{3}$$

Now letting $\varphi = \rho^{-n}\Phi$, we have

$$\rho^2(1 - \rho^2)(\rho^{-n}\Phi'' - 2n\rho^{-n-1}\Phi' + n(n + 1)\rho^{-n-2}\Phi)$$
$$+ \rho(1 - 2\rho^2)(\rho^{-n}\Phi' - n\rho^{-n-1}\Phi) - n^2\rho^{-n}\Phi = 0,$$

that is,

$$\rho(1 - \rho^2)\Phi'' - (2n - 1 - 2(n - 1)\rho^2)\Phi' - n(n - 1)\rho\Phi = 0. \qquad (4)$$

Again, let

$$\tau = 1 - \rho^2, \qquad (5)$$

we then obtain

$$\rho\tau\left(4\rho^2 \frac{d^2\Phi}{d\tau^2} - 2\frac{d\Phi}{d\tau}\right) + 2\rho[1 + 2(n - 1)\tau]\frac{d\Phi}{d\tau} - n(n - 1)\rho\Phi = 0,$$

that is,

$$4\tau(1 - \tau)\frac{d^2\Phi}{d\tau^2} + 2[1 + 2(n - 1)\tau - \tau]\frac{d\Phi}{d\tau} - n(n - 1)\Phi = 0, \qquad (6)$$

This is a hypergeometric differential equation, and in general its solutions are expressed in terms of the function

$$F(-\tfrac{1}{2}n, -\tfrac{1}{2}(n - 1), \tfrac{1}{2}, \tau).$$

However, for this special equation, we can directly prove the existence of the following two solutions:

$$(1 + \tau^{1/2})^n, \qquad (1 - \tau^{1/2})^n. \qquad (7)$$

For example, the first and second order derivatives of the former are

$$\frac{d}{d\tau}(1 + \tau^{1/2})^n = \frac{1}{2}n(1 + \tau^{1/2})^{n-1}\tau^{-1/2}$$

and

$$\frac{d^2}{d\tau^2}(1 + \tau^{1/2})^n = \frac{1}{4}n(n - 1)(1 + \tau^{1/2})^{n-2}\tau^{-1} - \frac{1}{4}n(1 + \tau^{1/2})^{n-1}\tau^{-3/2},$$

whereupon substitution into expression (6) yields

$$(1 - \tau)[n(n - 1)(1 + \tau^{1/2})^{n-2} - n(1 + \tau^{1/2})^{n-1}\tau^{-1/2}]$$
$$+ [1 + 2(n - 1)\tau - \tau]n(1 + \tau^{1/2})^{n-1}\tau^{-1/2} - n(n - 1)(1 + \tau^{1/2})^n$$
$$= (1 + \tau^{1/2})^{n-2}[n(n - 1)(1 - \tau)$$
$$- n(1 + \tau^{-1/2})(1 - \tau) + n(1 + 2(n - 1)\tau - \tau)$$
$$\times (1 + \tau^{-1/2}) - n(n - 1)(1 + \tau + 2\tau^{1/2})]$$
$$= 0.$$

However, the two solutions given by (7) are not always real. In order to study real solutions, we must use the following two real solutions:

$$P_n(\tau) = \tfrac{1}{2}[(1 + \tau^{1/2})^n + (1 - \tau^{1/2})^n] \qquad (8)$$

and

$$|\tau|^{1/2}Q_n(\tau), \tag{9}$$

where $Q_n(\tau) = 1/(2\tau^{1/2})((1 + \tau^{1/2})^n - (1 - \tau^{1/2})^n)$. By choosing

$$\tau^{1/2} = \begin{cases} |\tau|^{1/2}, & \text{for } \tau > 0, \\ i|\tau|^{1/2}, & \text{for } \tau < 0, \end{cases}$$

equation (D) has solutions of the form

$$\frac{P_n(\tau)}{\rho^n} \cos n\theta, \qquad \frac{P_n(\tau)}{\rho^n} \sin n\theta, \qquad |\tau|^{1/2} \frac{Q_n(\tau)}{\rho^n} \cos n\theta, \qquad |\tau|^{1/2} \frac{Q_n(\tau)}{\rho^n} \sin n\theta,$$

where $n = 1, 2, 3, \ldots$. These solutions all have singularities at the origin.

When $n = 0$, the above method requires further elaboration; from (1) we obtain:

$$\rho^2(1 - \rho^2)\varphi''(\rho) + \rho(1 - 2\rho^2)\varphi'(\rho) = 0, \qquad \psi''(\theta) = 0,$$

whence

$$\frac{\varphi''(\rho)}{\varphi'(\rho)} = \frac{2\rho^2 - 1}{\rho(1 - \rho^2)} = -\frac{1}{\rho} - \frac{1}{2}\frac{1}{1 + \rho} + \frac{1}{2}\frac{1}{1 - \rho},$$

so that

$$\varphi'(\rho) = -C_1 \frac{|\tau|^{1/2}}{\rho\tau},$$

and, in other words,

$$\varphi(\rho) = \begin{cases} C_1 \log \dfrac{1 + \sqrt{1 - \rho^2}}{\rho} + C_2, & \text{for } \rho < 1, \\[3mm] -C_1 \arccos \dfrac{1}{\rho} + C_2, & \text{for } \rho > 1 \end{cases}$$

(assuming $\phi(\rho)$ is continuous at $\rho = 1$), while $\psi(\theta) = C_3\theta + C_4$. However, since $u(\rho, \theta)$ is a function of θ with period 2π, $C_3 = 0$. Let

$$\sigma(\rho) = \begin{cases} \log \dfrac{1 + \sqrt{1 - \rho^2}}{\rho}, & \text{for } \rho < 1, \\[3mm] \arccos \dfrac{1}{\rho}, & \text{for } \rho > 1. \end{cases}$$

Then

$$\lim_{\rho \to 1 \pm 0} \frac{\sigma(\rho)}{|\tau|^{1/2}} = 1$$

and $\sigma(\rho)$ has a logarithmic singularity at the origin.

This suggests that differential equation (D) has a solution of the following form

$$u(\rho,\theta) = \frac{1}{2} a_0 + \sum_{n=1}^{\infty} (a_n \cos n\theta + b_n \sin n\theta) \times \frac{P_n(\tau)}{\rho^n}$$

$$+ c_0 \sigma(\rho) + |\tau|^{1/2} \times \sum_{n=1}^{\infty} (c_n \cos n\theta + d_n \sin n\theta) \frac{Q_n(\tau)}{\rho^n}. \qquad (F)$$

For the time being we shall put off the discussion of the convergence of (F), or whether or not (F) satisfies (D). Instead we will take an overview of the situation to see what the implications of (F) may be.

To begin with, we must first ascertain which class of functions are being discussed. We see from the form of (F), that when $\rho = 1$, we must weaken the conditions, i.e. $u(\rho,\theta)$ is differentiable at $\rho = 1$. Instead, we should assume the existence of

$$\lim_{\rho \to 1-0} \frac{u(\rho,\theta) - u(1,\theta)}{|\tau|^{1/2}} = \lim_{\rho \to 1+0} \frac{u(\rho,\theta) - u(1,\theta)}{|\tau|^{1/2}}. \qquad (10)$$

To be more precise, the class of functions which we are studying is the following:

Given a domain $\mathscr{D}$ which contains an arc of the unit circle; when $\rho \neq 1$, the function $u(\rho,\theta)$ is twice differentiable, but when $\rho = 1$, we assume that it satisfies (10). This class of functions is denoted by $\mathfrak{R}(\mathscr{D})$.

If we were to restrict ourselves to real analytic functions, then the second half of (F) would be nonexistent, and hence there would be no need to assume condition (10).

7.6 Some Examples

A great deal of research has been done in the area of partial differential equations of mixed type. Besides the Dirichlet problem or Neumann problem for elliptic equations and the Goursat problem for hyperbolic equations, there are in addition quite a few new problems. For example, problems on the type-changing curve, as well as problems of various mixed types; in other words, problems involving prescribing separate conditions in the hyperbolic and elliptic regions. However, since the main objective of this book is not to be comprehensive, but rather to consider typical cases, we will only cite two examples to illustrate the problem.

First of all, look at the case of the series (F). We have, by definition,

$$u(1,\theta) = \frac{1}{2} a_0 + \sum_{n=1}^{\infty} (a_n \cos n\theta + b_n \sin n\theta)$$

$$= \varphi(\theta).$$

From this we see that because c_n, d_n have not yet been determined in the series, we do not obtain a unique solution to equation (D) by having merely the value of $u(\rho, \theta)$ on the unit circle.

However, if we consider the class of functions $\mathfrak{R}(\mathscr{D})$, then we have

$$\lim_{\rho \to 1} (u(\rho, \theta) - u(1, \theta))/|\tau|^{1/2}$$

$$= c_0 \lim_{\rho \to 1} (\sigma(\rho))/|\tau|^{1/2} + \lim_{\tau \to 0} \sum_{n=1}^{\infty} (c_n \cos n\theta + d_n \sin n\theta) Q_n(\tau)$$

$$= c_0 + \sum_{n=1}^{\infty} n(c_n \cos n\theta + d_n \sin n\theta).$$

Denote the last expression by $\chi(\theta)$. This suggests:

Problem 1. Given $\varphi(\theta)$ and $\chi(\theta)$ on part of the unit circle in $\mathscr{D}$. Under what conditions is there a unique function in $\mathfrak{R}(\mathscr{D})$ satisfying

$$u(\rho, \theta)\big|_{\rho=1} = \varphi(\theta)$$

$$\lim_{\rho \to 1} \frac{u(\rho, \theta) - \varphi(\theta)}{|\tau|^{1/2}} = \chi(\theta).$$

Here we assume $\mathscr{D}$ does not contain the origin.

Let us now study a problem with conditions of mixed type.

Problem 2. On a characteristic line, we have

$$u(\rho, \theta)\big|_{x=1, y>0} = \tau(\theta), \qquad 0 < \theta < \frac{\pi}{2}. \tag{1}$$

On a diameter of the disc, we have

$$u(\rho, \theta)\big|_{|x|<1, y=0} = \psi(x). \tag{2}$$

Of course we have assumed that

$$\tau(0) = \psi(1).$$

Remark. We are striving for simplicity; otherwise the diameter could be replaced by any other curve.

We will now find the solution to equation (D) which satisfies conditions (1) and (2).

Outside the circle,

$$u(\rho, \theta) = f_1(\theta + \cos^{-1}(1/\rho)) + f_2(\theta - \cos^{-1}(1/\rho)),$$
$$f_2(0) = 0,$$

and condition (1) becomes

$$f_1(2\theta) = \tau(\theta), \qquad 0 < \theta < \frac{\pi}{2}.$$

Therefore,

$$u(\rho,\theta) = \tau(\tfrac{1}{2}(\theta + \cos^{-1}(1/\rho)) + f_2(\theta - \cos^{-1}(1/\rho)),$$
$$|x| < 1, \qquad \rho > 1,$$

and we easily obtain

$$\lim_{\rho \to 1+0} \left(\frac{\partial u}{\partial \theta} + \rho(\rho^2 - 1)^{1/2} \frac{\partial u}{\partial \rho} \right) = \tau'(\theta/2), \qquad 0 < \theta < \pi. \tag{3}$$

Consider now the case inside the circle; introduce the new variable

$$\lambda = \frac{1}{\rho} - \sqrt{\frac{1}{\rho^2} - 1}$$

and let

$$u(\rho,\theta) = U(\lambda,\theta).$$

As a result, equation (D) becomes

$$\frac{\partial^2 U}{\partial \lambda^2} + \frac{1}{\lambda} \frac{\partial U}{\partial \lambda} + \frac{1}{\lambda^2} \frac{\partial^2 U}{\partial \theta^2} = 0,$$

which is exactly the Laplace equation in polar coordinates, that is, $U(\lambda,\theta)$ is a harmonic function.

Since $u(\rho,\theta)$ belongs to the class of functions $\Re(\mathscr{D})$, it is not difficult to deduce from the definition that

$$\lim_{\rho \to 1+0} \left(\frac{\partial u}{\partial \theta} + \rho(\rho^2 - 1)^{1/2} \frac{\partial u}{\partial \rho} \right) = \lim_{\rho \to 1-0} \left(\frac{\partial u}{\partial \theta} + \rho(1 - \rho^2)^{1/2} \frac{\partial u}{\partial \rho} \right).$$

Hence from condition (3) we have

$$\lim_{\lambda \to 1-0} \left(\frac{\partial u}{\partial \theta} + \lambda \frac{\partial u}{\partial \lambda} \right) = \tau'(\theta/2), \qquad 0 \leqslant \theta < \pi. \tag{4}$$

Now let V be the harmonic conjugate of U, then

$$\left. \left(\frac{\partial u}{\partial \theta} + \frac{\partial V}{\partial \theta} \right) \right|_{\lambda=1} = \tau'(\theta/2),$$

that is,

$$(U + V)|_{\lambda=1} = 2\tau(\theta/2), \qquad 0 \leqslant \theta < \pi. \tag{5}$$

Again from (2), we can obtain

$$U|_{y=0} = \psi(x), \qquad -1 < x < 1. \tag{6}$$

Before solving for U and V, we introduce the following lemma:

Lemma. (*Keldeyšč-Sedov*) *Let s represent the boundary of the upper half disc S, and let $h(z)$ be an analytical function in S which is continuous on s, then*

$$h(z) = (1/\pi) \int_0^\pi \left[(1 - z^2) \Re h(e^{i\theta})/(1 - 2z \cos\theta + z^2) \right] d\theta$$

$$+ (1/\pi) \int_{-1}^1 \left[(1 - z^2) \Im h(t)/(t - z)(1 - tz) \right] dt. \tag{7}$$

PROOF. Let z be a point in S, then $1/z, \bar{z}, 1/\bar{z}$ are outside of S. From the Cauchy integral theorem we obtain

$$h(z) = \frac{1}{2\pi i} \int_S \frac{h(\zeta)}{\zeta - z} d\zeta = \frac{1}{2\pi} \int_0^\pi \frac{h(e^{i\theta})e^{i\theta}}{e^{i\theta} - z} d\theta + \frac{1}{2\pi i} \int_{-1}^1 \frac{h(t)}{t - z} dt \qquad (8)$$

$$0 = \frac{1}{2\pi i} \int_S \frac{h(\zeta)}{\zeta - \dfrac{1}{z}} d\zeta = \frac{1}{2\pi} \int_0^\pi \frac{h(e^{i\theta})e^{i\theta}}{e^{i\theta} - \dfrac{1}{z}} d\theta + \frac{1}{2\pi i} \int_{-1}^1 \frac{h(t)}{t - \dfrac{1}{z}} dt \qquad (9)$$

$$0 = \frac{1}{2\pi i} \int_S \frac{h(\zeta)}{\zeta - \bar{z}} d\zeta = \frac{1}{2\pi} \int_0^\pi \frac{h(e^{i\theta})e^{i\theta}}{e^{i\theta} - \bar{z}} d\theta + \frac{1}{2\pi i} \int_{-1}^1 \frac{h(t)}{t - \bar{z}} dt \qquad (10)$$

$$0 = \frac{1}{2\pi i} \int_S \frac{h(\zeta)}{\zeta - \dfrac{1}{\bar{z}}} d\zeta = \frac{1}{2\pi} \int_0^\pi \frac{h(e^{i\theta})e^{i\theta}}{e^{i\theta} - \dfrac{1}{\bar{z}}} d\theta + \frac{1}{2\pi i} \int_{-1}^1 \frac{h(t)}{t - \dfrac{1}{\bar{z}}} dt \qquad (11)$$

Subtracting (9) from (8) we have

$$h(z) = \frac{1}{2\pi} \int_0^\pi \frac{(1 - z^2)h(e^{i\theta})}{(1 - e^{-i\theta}z)(1 - e^{i\theta}z)} d\theta + \frac{1}{2\pi i} \int_{-1}^1 \frac{(1 - z^2)h(t)}{(t - z)(1 - tz)} dt,$$

and subtracting (11) from (10) and taking its conjugate, we have

$$0 = \frac{1}{2\pi} \int_0^\pi \frac{(1 - z^2)\overline{h(e^{i\theta})}}{(1 - e^{-i\theta}z)(1 - e^{i\theta}z)} d\theta - \frac{1}{2\pi i} \int_{-1}^1 \frac{(1 - z^2)\overline{h(t)}}{(t - z)(1 - tz)} dt.$$

The sum of the last two expressions is exactly (7). $\qquad\qquad\square$

Now let us use this lemma to find solutions to (5) and (6). Let

$$h(z) = i\left(\frac{1 + z}{1 - z}\right)^{1/2} f(z), \qquad f(z) = U + iV.$$

We have from (5)

$$\Re h(e^{i\theta}) = \Re(i\sqrt{i\,\text{ctg}\,(\theta/2)}(U + iV))_{\lambda - 1} = -\sqrt{\tfrac{1}{2}\,\text{ctg}\,(\theta/2)}(U + V)|_{\lambda = 1}$$
$$= -\sqrt{2\,\text{ctg}\,(\theta/2)}\tau(\theta/2),$$

$$\Im h(x) = \Im\left(i\sqrt{\frac{1 + x}{1 - x}}(U + iV)|_{y = 0}\right) = \sqrt{\frac{1 + x}{1 - x}}\psi(x)$$

and therefore our solution is

$$f(z) = \frac{1}{\pi i}(1 + z)^{1/2}(1 - z)^{3/2}\left(-\int_0^\pi \frac{\tau(\theta/2)}{1 - 2z\cos\theta + z^2}\right.$$

$$\left. \sqrt{2\,\text{ctg}\,\theta/2}\,d\theta + \int_{-1}^1 \frac{1}{(t - z)(1 - tz)}\sqrt{\frac{1 + t}{1 - t}}\psi(t)\,dt\right). \qquad (12)$$

From (12) we can derive $U(\lambda, \theta)$ and $u(\rho, \theta)$.

Similarly with (3), (4) and (5), we can obtain

$$\left(\frac{\partial u}{\partial \theta} - \rho\sqrt{\rho^2 - 1}\,\frac{\partial u}{\partial \rho}\right)\Bigg|_{\rho = 1} = f'(\theta/2), \qquad 0 \leqslant \theta < \pi$$

and

$$(U - V)|_{\lambda = 1} = 2f(\theta/2), \qquad 0 \leqslant \theta < \pi.$$

Thus

$$f_2(\theta) = (\tfrac{1}{2}(U - V))|_{\lambda = 1} = u(1, \theta) - \tfrac{1}{2}\tau(\theta), \qquad 0 \leqslant \theta < \pi.$$

Remark. When we use the Cauchy formula, we need to assume the convergence of

$$\sqrt{\frac{1 + z}{1 - z}}\,f(z)$$

on the boundary, but after getting (12), we can greatly weaken the assumptions on φ and ψ.

The foregoing proves the existence of a solution to Problem 2, but we have yet to determine whether or not it is unique. Since (12) has singularities at $z = 1, -1$, this question requires a deeper investigation.

What we have to discuss next is the following: whether or not there exists an $f(z)$ which is analytic in S, such that

$$U(\lambda, \theta)|_{|x| < 1,\ y = 0} = 0 \qquad\qquad\qquad (13)$$
$$(U + V)|_{\lambda = 1} = 0 \qquad\qquad\qquad\quad (14)$$
$$0 < \theta < \pi.$$

Let

$$g(w)iw^{1/2}f(z), \qquad w = \frac{1 - z}{1 + z}.$$

From (13) we know that when w is a positive real number, then $g(w)$ is real, and from (14) we know that when w lies on the positive imaginary axis, that is $w = iy$, $y > 0$, then

$$g(w) = i\,\frac{1 + i}{\sqrt{2}}\,y^{1/2}f\left(\frac{1 - iy}{1 + iy}\right) = \frac{-1 + i}{\sqrt{2}}\,y^{1/2}(U + iV)\Bigg|_{\lambda = 1}$$

$$= \frac{1}{\sqrt{2}}[-(U + V) + i(U - V)]_{\lambda = 1} = i\left(\frac{U - V}{2}\right)\Bigg|_{\lambda = 1}.$$

Thus $g(w)$ is pure imaginary. Because $g(w)$ is analytic within the first quadrant (not including $z = 0, \infty$), from the Schwarz reflection principle, we can extend $g(w)$ to be analytic in the fourth and second quadrants, as well as in the third, by defining:

$$g(x + iy) = g(x - iy) = g(-x + iy) = g(-x - iy).$$

Thus $g(w)$ is an even function, $g(-w) = g(w)$. Therefore we have the Laurent expansion

$$g(w) = \sum_{n=-\infty}^{\infty} b_n z^{2n} \quad \text{for } b_n \text{ real.}$$

Consequently,

$$f(z) = -w^{-1/2} g(w) = -i \sum_{n=-\infty}^{\infty} b_n w^{2n-1/2} = -i \sum_{n=-\infty}^{\infty} b_n \left(\frac{1-z}{1+z}\right)^{2n-1/2}.$$

Hence

$$U(\lambda, \theta) = \Re f(z) = \sum_{n=-\infty}^{\infty} b_n \Im \left(\frac{1-z}{1+z}\right)^{2n-1/2}.$$

Because we have

$$f'(z) = i \sum b_n (4n-1) \left(\frac{1-z}{1+z}\right)^{2n-3/2} \frac{1}{(1+z)^2},$$

it follows that if we add the condition

$$f'(z) = o(|1-z|)^{-3/2},$$

then $b_n = 0$ for $n \leqslant 0$. Thus

$$f'(z) = i \sum_{n=1}^{\infty} b_n (4n-1) \frac{(1-z)^{2n-3/2}}{(1+z)^{2n+1/2}}.$$

If we now add the condition

$$f'(z) = o(|1+z|)^{-5/2},$$

we have then $b_n = 0$ for $n \geqslant 1$. For this reason $f'(z) = 0$ and hence $f(z) = 0$.

So in order to arrive at a uniqueness theorem, we only need assume

$$\frac{\partial U}{\partial x}, \frac{\partial U}{\partial y} = o(|1-z|^{-3/2}) \quad \text{and} \quad o(|1+z|^{-5/2}),$$

for the following reason. From

$$\frac{\partial U}{\partial x}, \frac{\partial U}{\partial y} = o(|1-z|^{-3/2})$$

and the Cauchy-Riemann equation, we have

$$\frac{\partial V}{\partial x}, \frac{\partial V}{\partial y} = o(|1-z|^{-3/2}).$$

Therefore, when

$$1 - z = \rho e^{i\xi}, \qquad \frac{3\pi}{2} < \xi < 2\pi,$$

we have

$$f'(z) = o(|1 - z|^{-3/2}).$$

From the Schwarz reflection principle, when $0 \leqslant \xi \leqslant 2\pi$, we have

$$f'(z) = o(|1 - z|^{-3/2}).$$

We obtain

$$f'(z) = o(|1 + z|^{-5/2})$$

in exactly the same manner.

Remark. It is possible to establish uniqueness from the other conditions, such as

$$f'(z) = o(|1 - z|^{-7/2}) \quad \text{and} \quad o(|1 + z|^{-1/2}).$$

These conditions are better than those of Bitzadze.

7.7 Convergence of Series

Let us now consider the convergence of the series

$$u(\rho, \theta) = \frac{1}{2} a_0 + \sum_{n=1}^{\infty} (a_n \cos n\theta + b_n \sin n\theta) \frac{P_n(\tau)}{\rho^n}$$

$$+ c_0 \sigma(\rho) + |\tau|^{1/2} \sum_{n=1}^{\infty} (c_n \cos n\theta + d_n \sin n\theta) \frac{Q_n(\tau)}{\rho^n} \qquad \text{(F)}$$

and also whether or not it satisfies equation (D).

We first consider the case of the exterior of the unit circle, that is $\rho \geqslant 1$. Let

$$\rho = \frac{1}{\cos \eta}, \quad 0 \leqslant \eta < \frac{\pi}{2},$$

then

$$\frac{P_n(\tau)}{\rho^n} = \frac{1}{2} \left(\left(\frac{1 + \tau^{1/2}}{\rho} \right)^n + \left(\frac{1 - \tau^{1/2}}{\rho} \right)^n \right) = \cos n\eta,$$

$$|\tau|^{1/2} \frac{Q_n(\tau)}{\rho^n} = \frac{1}{2i} \left(\left(\frac{1 + \tau^{1/2}}{\rho} \right)^n - \left(\frac{1 - \tau^{1/2}}{\rho} \right) \right) = \sin n\eta,$$

$$\sigma(\rho) = \eta.$$

In this way, (F) becomes

$$u(\rho, \theta) = \frac{1}{2} a_0 + \sum_{n=1}^{\infty} (a_n \cos n\theta + b_n \sin n\theta) \cos n\eta$$

$$+ c_0 \eta + \sum_{n=1}^{\infty} (c_n \cos n\theta + d_n \sin n\theta) \sin n\eta. \qquad (1)$$

Since

$$\frac{\partial^2 u(\rho, \theta)}{\partial \theta^2} = - \sum_{n=1}^{\infty} n^2 (a_n \cos n\theta + b_n \sin n\theta) \cos n\eta$$

$$- \sum_{n=1}^{\infty} n^2 (c_n \cos n\theta + d_n \sin n\theta) \sin n\eta \qquad (2)$$

and

$$\frac{\partial^2 u(\rho, \theta)}{\partial \rho^2} = \sum_{n=1}^{\infty} (a_n \cos n\theta + b_n \sin n\theta) \left(-n^2 \cos n\eta \left(\frac{d\eta}{d\rho}\right)^2 \right.$$

$$\left. -n \sin n\eta \frac{d\eta}{d\rho} \right) + \sum_{n=1}^{\infty} (c_n \cos n\theta + d_n \sin n\theta)$$

$$\times \left[-n^2 \sin n\eta \left(\frac{d\eta}{d\rho}\right)^2 + n \cos n\eta \left(\frac{d\eta}{d\rho}\right) \right], \qquad (3)$$

it follows that if we assume

$$\sum_{n=1}^{\infty} (|a_n| + |b_n| + |c_n| + |d_n|) n^2 < \infty, \qquad (4)$$

then series (1), (2) and (3) all converge uniformly. Therefore, (F) satisfies (D) outside the unit circle.

Condition (4) obviously can be derived from the assumption that $\varphi(\theta)$ and $\chi(\theta)$ have, respectively, continuous fourth and third order derivatives. Equally obviously, it also can be derived from the assumption that $\varphi(\theta)$ and $\tau(\theta)$ both have continuous fourth order derivatives (or utilize other even weaker conditions from the theory of Fourier series).

Now we come to the case of the interior of the unit circle: we will prove the following assertions:

(i) If series (F) converges for $\rho = \rho_0$, $\theta = \theta_0$, then it also converges for $\rho > \rho_0$ when $\theta = \theta_0$.

(ii) If it converges on a θ set of positive measure when $\rho = \rho_0$, then it will converge when $\rho > \rho_0$ (moreover, it will converge uniformly on any finite subregion of this region).

The proof of (i) is extremely simple, for $P_n(\tau)/\rho^n$ and $Q_n(\tau)/\rho^n$ are decreasing functions of ρ, and from

$$\overline{\lim_{n \to \infty}} \left| (a_n \cos n\theta + b_n \sin n\theta) \frac{P_n(\tau_0)}{\rho_0^n} \right|^{1/n} \leqslant 1$$

and

$$\overline{\lim_{n \to \infty}} \left| (c_n \cos n\theta + d_n \sin n\theta) \frac{Q_n(\tau_0)}{\rho_0^n} \right|^{1/n} \leqslant 1,$$

we immediately obtain the fact that when $\rho > \rho_0$, there exists $\mu < 1$, such that

$$\overline{\lim_{n \to \infty}} \left| (a_n \cos n\theta + b_n \sin n\theta) \frac{P_n(\tau)}{\rho^n} \right|^{1/n} \leqslant \mu$$

and

$$\varlimsup_{n \to \infty} \left| (c_n \cos n\theta + d_n \sin n\theta) \frac{Q_n(\tau)}{\rho^n} \right|^{1/n} \leqslant \mu.$$

In order to prove (ii) we must make use of a lemma: Given a θ point set of positive measure on which the following holds:

$$\varlimsup_{n \to \infty} |a_n \cos n\theta + b_n \sin n\theta|^{1/n} = v,$$

then

$$\varlimsup_{n \to \infty} |a_n^2 + b_n^2|^{1/2n} = v.$$

This lemma is known (it can be derived using the theory of uniform distribution in number theory) (Lusin and Steinhaus, see Zygmund, *Trigonometric Series*, p. 131 and p. 269).

Armed with this lemma, we can at once use the preceding method to prove (ii) as well the condition

$$a_n, b_n, c_n, d_n = O(\rho_0^n).$$

If this last condition is satisfied, condition (4) naturally follows.

To sum up, if on a circle centered at the origin and lying within the unit disc, there exists a set of positive measure on which (F) converges, then it also converges everywhere outside this circle and satisfies equation (D).

7.8 Functions Without Singularities Inside the Unit Circle (Analogues of Holomorphic Functions)

Question. Given a function on a characteristic line, does it determine a function which everywhere satisfies (D)? Is it unique?

Since the characteristic lines form a set transitive under Γ, it makes no difference which one we study. Thus there is no harm in choosing $x = 1$, that is

$$\rho = \frac{1}{\cos \theta}, \qquad |\theta| \leqslant \frac{\pi}{2}.$$

Then a more precise statement of the question is: assuming that

$$u(\rho, \theta)\big|_{x=1} = \tau(\theta),$$

we wish to find a $u(\rho, \theta)$ which satisfies (D).

(1) Existence. Because of objective requirements, we will assume that $\tau(\theta)$ has second order derivatives (if $u(\rho, \theta)$ does not possess a second order partial derivative with respect to θ, then we will have to discuss generalized

solutions) and that is a function of period π. Let

$$\tau(\theta) = \frac{1}{2} p_0 + \sum_{n=1}^{\infty} (p_n \cos 2n\theta + q_n \sin 2n\theta)$$

be the Fourier expansion of $\tau(\theta)$; the series

$$\sum_{n=1}^{\infty} (|p_n| + |q_n|)$$

clearly converges.

Conclusion: Let

$$\alpha_n = p_n + q_n, \qquad \beta_n = q_n - p_n, \qquad \alpha_0 = p_0 - 2 \sum_{n=1}^{\infty} q_n,$$

then the function

$$u(\rho, \theta) = \frac{1}{2} \alpha_0 + \sum_{n=1}^{\infty} (\alpha_n \cos n\theta + \beta_n \sin n\theta) \frac{P_n(\tau) - |\tau|^{1/2} Q_n(\tau)}{\rho^n} \tag{1}$$

satisfies our requirement.

The proof goes on as follows. On the characteristic line, we have

$$\frac{P_n(\tau)}{\rho^n} = \frac{1}{2} \left[\left(\frac{1}{\rho} + i \sqrt{1 - \frac{1}{\rho^2}} \right)^n + \left(\frac{1}{\rho} - i \sqrt{1 - \frac{1}{\rho^2}} \right)^n \right]$$

$$= \cos n\theta,$$

$$|\tau|^{1/2} \frac{Q_n(\tau)}{\rho^n} = \sin n\theta.$$

Therefore,

$$u(\rho, \theta)|_{x=1} = \frac{1}{2} \alpha_0 + \sum_{n=1}^{\infty} (\alpha_n \cos n\theta + \beta_n \sin n\theta)(\cos n\theta + \sin n\theta)$$

$$= \frac{1}{2} \alpha_0 + \frac{1}{2} \sum_{n=1}^{\infty} [\alpha_n(\cos 2n\theta + 1 + \sin 2n\theta)$$

$$+ \beta_n(\sin 2n\theta - \cos 2n\theta + 1)]$$

$$= \frac{1}{2} p_0 + \sum_{n=1}^{\infty} (p_n \cos 2n\theta + q_n \sin 2n\theta)$$

$$= \tau(\theta).$$

Next, outside the unit circle, (1) is everywhere convergent and it satisfies (D). Outside the circle, let

$$\rho = \frac{1}{\cos \eta}, \qquad 0 \leqslant \eta < \frac{\pi}{2},$$

then

$$\frac{P_n(\tau)}{\rho^n} = \cos n\eta, \qquad |\tau|^{1/2}\frac{Q_n(\tau)}{\rho^n} = \sin n\eta.$$

Hence (1) becomes

$$u(\rho,\theta) = \frac{1}{2}\alpha_0 + \sum_{n=1}^{\infty}(\alpha_n \cos n\theta + \beta_n \sin n\theta)(\cos n\eta + \sin n\eta),$$

and since $\sum_{n=1}^{\infty}(|\alpha_n| + |\beta_n|) < \infty$, this series converges.

If we assume

$$\sum_{n=1}^{\infty} n^2(|p_n| + |q_n|) < \infty, \tag{2}$$

then

$$\frac{\partial^2 u(\rho,\theta)}{\partial\theta^2} = -\sum_{n=1}^{\infty} n^2(\alpha_n \cos n\theta + \beta_n \sin n\theta)(\cos n\eta + \sin n\eta)$$

and

$$\rho(\rho^2-1)^{1/2}\frac{\partial u(\rho,\theta)}{\partial\rho} = \rho(\rho^2-1)^{1/2}\frac{\partial u(\rho,\theta)}{\partial\eta}\frac{\partial\eta}{\partial\rho} = \frac{\partial u(\rho,\theta)}{\partial\eta}$$

$$= \sum_{n=1}^{\infty} n(\alpha_n \cos n\theta + \beta_n \sin n\theta)(-\sin n\eta + \cos n\eta).$$

From this we know that (1) satisfies equation (D). Of course we have had to add some assumptions, such as: $\tau(\theta)$ has a continuous fourth order derivative. However, if this assumption is not satisfied, then we may look at (1) as a generalized solution of equation (D) in the region of hyperbolicity.

Finally, inside the unit circle, we introduce the new variable

$$\lambda = \frac{1}{\rho} - \sqrt{\frac{1}{\rho^2}-1} = \frac{1-\tau^{1/2}}{\rho} = \frac{\rho}{1+\tau^{1/2}}.$$

Since

$$\frac{d\lambda}{d\rho} = \frac{1-\sqrt{1-\rho^2}}{\rho^2\sqrt{1-\rho^2}} \geqslant 0,$$

when ρ goes from 0 to 1, λ is monotone increasing from 0 to 1. This then gives us

$$\frac{P_n(\tau)}{\rho^n} = \frac{1}{2}\left(\left(\frac{1+\tau^{1/2}}{\rho}\right)^n + \left(\frac{1-\tau^{1/2}}{\rho}\right)^2\right) = \frac{1}{2}(\lambda^n + \lambda^{-n}),$$

$$|\tau|^{1/2}\frac{Q_n(\tau)}{\rho^n} = \frac{1}{2}(-\lambda^n + \lambda^{-n}).$$

Hence, series (1) becomes

$$u(\rho, \theta) = \frac{1}{2} \alpha_0 + \sum_{n=1}^{\infty} (\alpha_n \cos n\theta + \beta_n \sin n\theta)\lambda^{-n},$$

and it is a harmonic function of (λ^{-1}, θ) and converges everywhere inside the circle. Therefore (1) satisfies (D) inside the circle.

(2) Uniqueness. We need to prove that $u(\rho, \theta) \equiv 0$ is the unique solution satisfying

$$u(\rho, \theta)|_{x=1} = 0, \qquad -\frac{\pi}{2} < \theta < \frac{\pi}{2},$$

and that it has no singularities inside the unit circle.

Outside the unit circle we have the general solution

$$u(\rho, \theta) = F_1(\theta + \cos^{-1}(1/\rho)) + F_2(\theta - \cos^{-1}(1/\rho)),$$

$$F_2(0) = 0,$$

so that for this solution,

$$u(\rho, \theta)|_{x=1} = F_1(2\theta) = 0.$$

Hence, outside the circle,

$$u(\rho, \theta) = F_2(\theta - \cos^{-1}(1/\rho)), \qquad F_2(0) = 0. \qquad (3)$$

By treating the case of inside the circle in a manner similar to the one we used to study Problem 2 of §6, we obtain

$$\left(\frac{\partial U}{\partial \theta} + \lambda \frac{\partial U}{\partial \lambda} \right)\Bigg|_{\lambda=1} = 0.$$

Suppose U is the real part of an analytic function $f(z)$ of a complex variable in the unit disc, and V is the imaginary part, such that $f(z)$ is continuous up to the boundary and satisfies $f(1) = 0$. Then

$$(U + V)|_{\lambda=1} = 0,$$

and, as a result, the function

$$f(z) = U(\lambda, \theta) = iV(\lambda, \theta)$$

transforms the unit circle $|z| = 1$ into the straight line

$$U + V = 0.$$

This utilizes the Schwarz reflection principle (as in Question 2 of §6). Under certain additional conditions, the analyticity of the function $f(z)$ can be extended to the whole plane (including points at infinity), from which it follows that $U = 0$.

7.9 Functions Having Logarithmic Singularities Inside the Circle

Let us begin with

$$\log \lambda.$$

It is a harmonic function which has a logarithmic singularity at the origin. In terms of polar coordinates (ρ, θ), we obtain a fundamental solution to (D)

$$\sigma(\rho) = \begin{cases} \log\left(\dfrac{1}{\rho} - \sqrt{\dfrac{1}{\rho^2} - 1}\right), & \text{for } \rho \leqslant 1, \\[2ex] \cos^{-1}\dfrac{1}{\rho}, & \text{for } \rho \geqslant 1 \end{cases}$$

or

$$\sigma(x, y) = \begin{cases} \log\left(\dfrac{1 - \sqrt{1 - x^2 - y^2}}{\sqrt{x^2 + y^2}}\right), & \text{for } x^2 + y^2 \leqslant 1, \\[2ex] \cos^{-1}\dfrac{1}{\sqrt{x^2 + y^2}}, & \text{for } x^2 + y^2 \geqslant 1. \end{cases}$$

We will now make use of the action of the group Γ. Let

$$x = f(x', y', a, b), \qquad y = g(x', y', a, b)$$

represent a transformation by an element of Γ which takes (a, b) into the origin; in this way

$$\sigma(x, y) = \sigma(f(x', y', a, b), g(x', y', a, b))$$
$$= \sigma_{a,b}(x', y').$$

Because of the invariance of the partial differential equation, we know that $\sigma_{a,b}$ is still a solution to (D).

Let $\mu(a, b)$ be any distribution function, then the function

$$F(x, y) = \iint\limits_{a^2 + b^2 \leqslant 1} \sigma_{a,b}(x, y)\, d\mu(a, b)$$

is still a solution to equation (D).

Question. What kind of functions is $F(1, y)$, that is, from which class of functions $\{\varphi(y)\}$ is it; can we find for every such $\varphi(y)$ a $\mu(a, b)$ such that

$$\varphi(y) = \iint\limits_{a^2 + b^2 \leqslant 1} \sigma_{a,b}(1, y)\, d\mu(a, b)?$$

To put it more concretely, let

$$x_1 = (x \cos \alpha + y \sin \alpha - \mu)/(1 - \mu x \cos \alpha - \mu y \sin \alpha),$$
$$y_1 = \sqrt{1 - \mu^2}(-x \sin \alpha + y \cos \alpha)/(1 - \mu x \cos \alpha - \mu y \sin \alpha)$$

be a transformation belonging to Γ which takes the point $(\mu \cos \alpha, \mu \sin \alpha)$ inside the circle into $(0,0)$. Hence we have the function

$$\sigma_{\mu \cos \alpha, \mu \sin \alpha}(x, y) = \sigma\left(\frac{\sqrt{(x \cos \alpha + y \sin \alpha - \mu)^2 + (1 - \mu^2)(-x \sin \alpha + y \cos \alpha)^2}}{1 - \mu x \cos \alpha - \mu y \sin \alpha}\right)$$

$$= \sigma\left(\frac{\sqrt{(1 - \mu^2)(x^2 + y^2 - 1) + (1 - \mu(x \cos \alpha + y \sin \alpha))^2}}{1 - \mu(x \cos \alpha + y \sin \alpha)}\right)$$

$$= \sigma\left(\frac{\sqrt{(1 - \mu^2)(\rho^2 - 1) + (1 - \mu\rho \cos(\alpha - \theta))^2}}{1 - \mu\rho \cos(\alpha - \theta)}\right)$$

$$(x = \rho \cos \theta, \, y = \rho \sin \theta).$$

When $x = 1$, $y = \tan \theta$,

$$\varphi(\tan \theta) = \int_0^1 \int_0^{2\pi} \cos^{-1} \frac{\cos \theta - \mu \cos(\alpha - \theta)}{\sqrt{(1 - \mu^2) \sin^2 \theta + (\cos \theta - \mu \cos(\alpha - \theta))^2}} \, dq\,(\alpha, \mu).$$

The question then becomes: for what kind of $\varphi(\tan \theta)$ can we always find a function of bounded variation, $q(\alpha, \mu)^2$, so that the above holds?

7.10 The Poisson Formula

Given a general transformation in Γ:

$$x' = \frac{a_1 x + b_1 y + c_1}{a_3 x + b_3 y + c_3}, \qquad y' = \frac{a_2 x + b_2 y + c_2}{a_3 x + b_3 y + c_3} \tag{1}$$

let

$$x = \cos \theta, \qquad y = \sin \theta, \qquad x' = \cos \theta', \qquad y' = \sin \theta'.$$

Then we have

$$\cos \theta' = \frac{a_1 \cos \theta + b_1 \sin \theta + c_1}{a_3 \cos \theta + b_3 \sin \theta + c_3},$$

$$\sin \theta' = \frac{a_2 \cos \theta + b_2 \sin \theta + c_2}{a_3 \cos \theta + b_3 \sin \theta + c_3},$$

so that

$$\tan \theta' = \frac{a_2 \cos \theta + b_2 \sin \theta + c_2}{a_1 \cos \theta + b_1 \sin \theta + c_1}. \tag{2}$$

Thus, from the relationship involving a, b, and c in §1, we obtain

$$\frac{d\theta'}{d\theta} = ((a_1 \cos \theta + b_1 \sin \theta + c_1)(-a_2 \sin \theta + b_2 \cos \theta)$$

$$- (a_2 \cos \theta + b_2 \sin \theta + c_2)(-a_1 \sin \theta + b_1 \cos \theta))/((a_2 \cos \theta + b_2 \sin \theta + c_2)^2 + (a_1 \cos \theta + b_1 \sin \theta + c_1)^2)$$

$$= (a_1 b_2 - a_2 b_1 + (-c_1 a_2 + c_2 a_1) \sin \theta$$

$$+ (c_1 b_2 - c_2 b_1) \cos \theta)/((1 + a_3^2) \cos^2 \theta + 2a_3 b_3 \cos \theta \sin \theta$$

$$+ (1 + b_3^2) \sin^2 \theta + 2a_3 c_3 \cos \theta + 2b_3 c_3 \sin \theta + c_3^2 - 1)$$

$$= \frac{1}{a_3 \cos \theta + b_3 \sin \theta + c_3}, \tag{3}$$

whence

$$1 = \frac{1}{2\pi} \int_0^{2\pi} d\theta' = \frac{1}{2\pi} \int_0^{2\pi} \frac{d\theta}{|a_3 \cos \theta + b_3 \sin \theta + c_3|}. \tag{4}$$

Let us assume that (1) takes (ξ, η) into the origin, that is

$$a_1 \xi + b_1 \eta + c_1 = 0, \qquad a_2 \xi + b_2 \eta + c_2 = 0, \tag{5}$$

where (ξ, η) is inside the unit circle. Then from (5) we know that

$$\xi = -\frac{a_3}{c_3}, \qquad \eta = -\frac{b_3}{c_3},$$

and from

$$a_3^2 + b_3^2 - c_3^2 = -1,$$

we have

$$c_3^2(\xi^2 + \eta^2 - 1) = -1.$$

Substitution into (3) yields

$$\frac{d\theta'}{d\theta} = \pm \frac{\sqrt{1 - \xi^2 - \eta^2}}{1 - \xi \cos \theta - \eta \sin \theta}.$$

Transforming (ξ, η) into (x, y), we then get from (4) the fact that for any point (x, y) inside the circle, we always have

$$1 = \frac{1}{2\pi} \int_0^{2\pi} \frac{\sqrt{1 - x^2 - y^2}}{1 - x \cos \theta - y \sin \theta} d\theta. \tag{6}$$

(This is true since $|x \cos \theta + y \sin \theta| < 1$.) If we use polar coordinates, we have

$$1 = \frac{1}{2\pi} \int_0^{2\pi} \frac{\sqrt{1 - \rho^2}}{1 - \rho \cos(\theta - \psi)} d\theta. \tag{7}$$

If (x, y) lies outside the circle, i.e. when $\rho > 1$, then we claim:

$$\int_0^{2\pi} \frac{d\theta}{1 - \rho \cos (\theta - \psi)} = 0. \tag{8}$$

The proof of this claim is simple. Since the indefinite integral of $1/(1 - \rho \cos \theta)$ is equal to

$$\frac{1}{\rho^2 - 1} \log \left| \frac{\sqrt{\rho^2 - 1} \tan \frac{1}{2}\theta + 1 - \rho}{\sqrt{\rho^2 - 1} \tan \frac{1}{2}\theta - 1 + \rho} \right|,$$

we thus have

$$\int_0^\pi \frac{d\theta}{1 - \rho \cos \theta} = \lim_{\varepsilon \to 0} \left(\int_0^{\cos^{-1} 1/\rho - \varepsilon} + \int_{\cos^{-1} 1/\rho + \varepsilon}^\pi \right) \frac{d\theta}{1 - \rho \cos \theta}$$

$$= \frac{1}{\rho^2 - 1} \lim_{\varepsilon \to 0} \left(\log \left| \frac{\sqrt{\rho^2 - 1} \tan \frac{1}{2}\theta + 1 - \rho}{\sqrt{\rho^2 - 1} \tan \frac{1}{2}\theta - 1 - \rho} \right| \Big|_{\cos^{-1} 1/\rho + \varepsilon}^{\cos^{-1} 1/\rho - \varepsilon} \right)$$

$$= \frac{1}{\sqrt{\rho^2 - 1}} \lim_{\varepsilon \to 0} \log \frac{\sqrt{\rho^2 - 1} \tan \frac{1}{2} \left(\cos^{-1} \dfrac{1}{\rho} - \varepsilon \right) + 1 - \rho}{\sqrt{\rho^2 - 1} \tan \frac{1}{2} \cos^{-1} \dfrac{1}{\rho} + \varepsilon + 1 - \rho}$$

$$= 0.$$

Introduce a variable

$$P(\rho, \theta - \psi) = \frac{|\tau|^{1/2}}{1 - \rho \cos (\theta - \psi)}, \qquad \tau = 1 - \rho^2,$$

and call it the Poisson kernel. This function has the following four properties:

(1) Consider $P(\rho, \theta - \psi)$ as a function of polar coordinates (ρ, θ), then it satisfies (D) regardless of whether it lies inside the circle or outside the circle.
(2) Except at the point $\theta = \psi$, this function is everywhere zero on the circle.
(3) Except along a straight line (one of the characteristic lines, $\rho \cos (\theta - \psi) = 1$), where it is infinite, this function is bounded on the whole plane.
(4) We have the relation

$$\frac{1}{2\pi} \int_0^{2\pi} P(\rho, \theta) \, d\theta = \begin{cases} 0, & \text{for } \rho > 1, \\ 1, & \text{for } \rho < 1. \end{cases}$$

If we are given

$$u(\rho, \theta)\big|_{\rho = 1} = \alpha(\theta),$$

and construct the function

$$u(\rho, \theta) = \frac{1}{2\pi} \int_0^{2\pi} P(\rho, \theta - \psi)\alpha(\psi) \, d\psi, \tag{9}$$

then this function will satisfy the partial differential equation (D) inside the circle.

Outside the circle, let

$$\rho = \frac{1}{\cos \eta}, \qquad 0 \leqslant \eta \leqslant \frac{\pi}{2},$$

then

$$P(\eta, \theta - \psi) = \frac{\sin \eta}{\cos \eta - \cos(\theta - \psi)}$$

$$= -\frac{\sin \eta}{2 \sin \tfrac{1}{2}(\eta + \theta - \psi) \sin \tfrac{1}{2}(\eta - \theta + \psi)}$$

$$= -\tfrac{1}{2}(\tan \tfrac{1}{2}(\eta + \theta - \psi) + \tan \tfrac{1}{2}(\eta - \theta + \psi)).$$

Thus we get

$$u(\rho, \theta) = F_1(\eta + \theta) + F_2(\eta - \theta),$$

where

$$F_1(\gamma) = -\frac{1}{4\pi} \int_0^{2\pi} \tan \frac{1}{2}(\gamma - \psi)\alpha(\psi)\, d\psi,$$

$$F_2(\gamma) = -\frac{1}{4\pi} \int_0^{2\pi} \tan \frac{1}{2}(\gamma + \psi)\alpha(\psi)\, d\psi.$$

These singular integrals may not converge, and even if they are convergent, they may not be differentiable. However, we can deal with (9) in this case as a generalized solution to (D) in the region of hyperbolicity.

If we were to add conditions to $\alpha(\psi)$ to guarantee that F_1 and F_2 exist, and be twice differentiable, then such a $u(\rho, \theta)$ would be a solution to equation (D).

7.11 Functions with Prescribed Values on the Type-Changing Curve

Suppose we are given

$$u(\rho, \theta)\big|_{\rho=1} = F(\theta) \tag{1}$$

and

$$\lim_{\rho \to 1 \pm 0} \frac{u(\rho, \theta) - F(\theta)}{|\tau|^{1/2}} = G(\theta), \tag{2}$$

where $F(\theta)$ and $G(\theta)$ are real analytic functions for $\alpha < \theta < \beta$ and we wish to determine the function $u(\rho, \theta)$ which satisfies (D).

Inside the circle, let

$$u(\rho, \theta) = U(\lambda, \theta), \tag{3}$$

then

$$U(1, \theta) = F(\theta), \qquad \frac{\partial U}{\partial \lambda}\bigg|_{\lambda = 1} = -G(\theta),$$

This is a Cauchy problem and its solution is as follows:

Let us first consider the harmonic function which satisfies

$$U_1(1, \theta) = F(\theta), \qquad \frac{\partial U_1}{\partial \lambda}\bigg|_{\lambda = 1} = 0. \tag{4}$$

The function we are looking for is

$$U_1(\lambda, \theta) = \sum_{n=0}^{\infty} (-1)^n \frac{(\log \lambda)^{2n}}{(2n)!} F^{(2n)}(\theta), \tag{5}$$

for the following reasons:

(i) $U_1(1, \theta) = F(\theta)$,

(ii) $\lambda \dfrac{\partial U_1}{\partial \lambda}\bigg|_{\lambda = 1} = \sum_{n=1}^{\infty} (-1)^n \dfrac{(\log \lambda)^{2n-1}}{(2n-1)!} F^{(2n)}(\theta)\bigg|_{\lambda - 1} = 0,$

(iii) $\lambda \dfrac{\partial}{\partial \lambda}\left(\lambda \dfrac{\partial U_1}{\partial \lambda}\right) = \sum_{n=1}^{\infty} (-1)^n \dfrac{(\log \lambda)^{2n-2}}{(2n-2)!} F^{(2n)}(\theta)$

$$= -\frac{\partial^2}{\partial \theta^2}\left(\sum_{m=0}^{\infty} (-1)^m \frac{(\log \lambda)^{2m}}{(2m)!} F^{(2m)}(\theta)\right).$$

Since

$$F(\theta + x) = \sum_{n=0}^{\infty} \frac{F^{(n)}(\theta)}{n!} x^n,$$

we know that

$$F(\theta + i \log \lambda) + F(\theta - i \log \lambda) = \sum_{n=0}^{\infty} \frac{F^{(n)}(\theta)}{n!}(i^n + (-i)^n)(\log \lambda)^n,$$

and from (5) we get

$$U_1(\lambda, \theta) = \tfrac{1}{2}(F(\theta + i \log \lambda) + F(\theta - i \log \lambda)) \tag{6}$$

as desired.

Using the same method, we find that the harmonic function satisfying

$$U_2(1, \theta) = 0, \qquad \frac{\partial U_2}{\partial \lambda}\bigg|_{\lambda = 1} = G(\theta)$$

is

$$U_2(\lambda, \theta) = \sum_{n=0}^{\infty} (-1)^n \frac{(\log \lambda)^{2n+1}}{(2n+1)!} G^{(2n)}(\theta)$$

$$= \frac{1}{2i}(G_1(\theta + i \log \lambda) - G_1(\theta - i \log \lambda)),$$

where

$$G_1(\theta) = \int_0^\theta G(t)\,dt.$$

Hence

$$U(\lambda, \theta) = U_1(\lambda, \theta) - U_2(\lambda, \theta) = \frac{1}{2}(F(\theta + i\log\lambda)$$

$$+ F(\theta - i\log\lambda)) - \frac{1}{2i}(G_1(\theta + i\log\lambda) - G_1(\theta - i\log\lambda)).$$

Returning to our original notation, we have inside the circle,

$$\sigma(\rho) = \log\left(\frac{1}{\rho} + \sqrt{\frac{1}{\rho^2} - 1}\right)$$

$$= -\log\left(\frac{1}{\rho} - \sqrt{\frac{1}{\rho^2} - 1}\right)$$

$$= -\log\lambda.$$

Therefore, when $\rho \leqslant 1$,

$$u(\rho, \theta) = \frac{1}{2}\left[F(\theta + i\sigma(\rho)) + F(\theta - i\sigma(\rho))\right]$$

$$+ \frac{1}{2i}\left[G_1(\theta + i\sigma(\rho)) - G_1(\theta - i\sigma(\rho))\right]. \tag{7}$$

For the outside of the circle,

$$u(\rho, \theta) = \frac{1}{2}\left[F\left(\theta + \cos^{-1}\frac{1}{\rho}\right) + F\left(\theta - \cos^{-1}\frac{1}{\rho}\right)\right]$$

$$+ \frac{1}{2}\left[G_1\left(\theta + \cos^{-1}\frac{1}{\rho}\right) - G_1\left(\theta - \cos^{-1}\frac{1}{\rho}\right)\right]. \tag{8}$$

It is very easily proved that

$$u(1, \theta) = F(\theta),$$

and

$$\lim_{\rho \to \pm 1} \frac{u(\rho, \theta) - u(1, \theta)}{|\tau|^{1/2}} = G(\theta).$$

Consequently, (7) and (8) yield a solution to our problem.

Let us look at the region outside the unit circle where this solution is applicable. Since we must have

$$\alpha \leqslant \theta - \cos^{-1}\frac{1}{\rho} < \theta + \cos^{-1}\frac{1}{\rho} \leqslant \beta,$$

we know that

$$\rho \cos(\theta - \alpha) \leqslant 1 \quad \text{and} \quad \rho \cos(\beta - \theta) \leqslant 1,$$

and this is the region between two tangents to the unit circle, one tangent to the circle at $\theta = \alpha$ and the other at $\theta = \beta$.

Suppose $\alpha = 0$; let us look at the restriction of this function to the line $x = 1$, i.e. $\rho = 1/(\cos\theta)$, then

$$u\left(\frac{1}{\cos\theta}, \theta\right) = \frac{1}{2}(F(2\theta) + F(0)) + \frac{1}{2}(G_1(2\theta) - G_1(0)), \qquad 0 < \theta < \beta.$$

7.12 Functions Vanishing on a Characteristic Line

Consider again the class of functions which satisfy

$$u(\rho, \theta)|_{x=1} = 0. \tag{1}$$

Since outside the circle we have

$$u(\rho, \theta) = F_1\left(\theta + \cos^{-1}\frac{1}{\rho}\right) + F_2\left(\theta - \cos^{-1}\frac{1}{\rho}\right),$$

$$F_2(0) = 0,$$

we immediately get

$$u(\rho, \theta)|_{x=1} = F_1(2\theta) = 0.$$

Hence

$$u(\rho, \theta) = F_2\left(\theta - \cos^{-1}\frac{1}{\rho}\right), \qquad F_2(0) = 0.$$

From

$$\lim_{\rho \to 1} \frac{u(\rho, \theta) - u(1, \theta)}{|\tau|^{1/2}} = \lim_{\rho \to 1} \frac{F_2\left(\theta - \cos^{-1}\frac{1}{\rho}\right)}{|\tau|^{1/2}} = F_2'(\theta),$$

letting $u(\rho, \theta) = U(\lambda, \theta)$, then the function which satisfies (1) always satisfies

$$U(1, \theta) = F_2(\theta), \qquad \frac{\partial}{\partial \lambda} U(\lambda, \theta)|_{\lambda=1} = -F_2'(\theta). \tag{2}$$

Suppose $u(\rho, \theta)$ is well-defined on a subregion of the unit disc such that part of its boundary is an arc on the unit circle. Denote the corresponding region in the (λ, θ) plane by D, then part of the boundary of D is also an arc on the unit circle, to be denoted by γ; γ contains the point $(1, 0)$. The other part of the boundary of D is a curve, to be denoted by Γ.

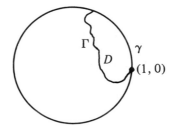

Inside D, $U(\lambda, \theta)$ is the real part of an analytic function.
Write
$$Q = f(z) = U(\lambda, \theta) + iV(\lambda, \theta).$$
Since
$$\lambda \frac{\partial}{\partial \lambda} U(\lambda, \theta) = \frac{\partial}{\partial \theta} V(\lambda, \theta),$$
thus
$$V(1, \theta) = -F_2(\theta) + C,$$
where C is a constant. We may assume that it equals 0. The conformal
transformation
$$Z = f(z)$$
takes D into D^* and takes the circular arc γ on the boundary of D into the
straight line
$$U + V = 0.$$

If we assume that this conformal transformation is one-to-one, then from
the Schwarz reflection principle, $f(z)$ can be analytically continued to the
exterior of the unit circle. Let Γ' be the curve gotten by reflecting Γ across the
circle, then $f(z)$ is defined in the region enclosed by Γ' and Γ. The real part of
$f(z)$ is already fixed on Γ, and on Γ', U is already defined. From its values on
Γ and Γ', we conclude that U not only exists, but is also unique. (The details
are similar to those in the work of Bitzadze, and make use of the Keldeyšč-
Sedov formula.)

CHAPTER 8
Formal Fourier Series and Generalized Functions

8.1 Formal Fourier Series

We have already more than once mentioned the result that, given a converging Fourier series

$$a_0 + \sum_{n=1}^{\infty} (a_n \cos n\theta + b_n \sin n\theta), \tag{1}$$

there exists on the unit disc a harmonic function

$$a_0 + \sum_{n=1}^{\infty} (a_n \cos n\theta + b_n \sin n\theta)\rho^n, \qquad 0 \leqslant \rho < 1. \tag{2}$$

Conversely, suppose (2) converges for all ρ satisfying $0 \leqslant \rho < 1$, we then say that (1) defines a *generalized function*. Since a harmonic function is infinitely differentiable, then so is a generalized function.

Although it may appear to be simple, the generalized functions defined in this manner are of greater scope than those of L. Schwarz.[1] One can achieve even greater generality by considering formal Fourier series, as we now do.

A *formal Fourier series* is a formal sum

$$\sum_{n=-\infty}^{\infty} a_n e^{in\theta};$$

it is not required to be convergent or summable in any sense. A *generalized function* is by definition such a formal Fourier series; we denote it by $u(\theta)$.

---

[1] Translator's note: these remarks are understood in the context of $\mathbb{R}^2$, of course.

From now on we will use $\sum_n$ to represent $\sum_{n=-\infty}^{\infty}$, and $\sum_n'$ to represent $\sum_n$ without the term $n = 0$. Letting

$$v(\theta) = \sum_{n=-\infty}^{\infty} b_n e^{in\theta},$$

then for any two complex number λ, μ, the formal Fourier series

$$\lambda u(\theta) + \mu v(\theta) = \sum_{n=-\infty}^{\infty} (\lambda a_n + \mu b_n)e^{in\theta}$$

is still a generalized function. Thus the generalized functions form a linear set.

The product of two generalized functions is, in general, not definable, the reason for this being that

$$\sum_{l+m=n} a_l b_m$$

is not necessarily convergent.

However, if

$$\sum_n a_n \bar{b}_n$$

converges (or is summable in a certain sense), then this value is called the *scalar* or *inner product* of the two generalized functions $u(\theta)$ and $\overline{v(\theta)}$, and is denoted by $(u(\theta), \overline{v(\theta)})$.

It is obvious that

$$(u(\theta), \overline{v(\theta)}) = \overline{(v(\theta), \overline{u(\theta)})},$$

$$(\lambda u_1(\theta) + \mu u_2(\theta), \overline{v(\theta)}) = \lambda(u_1(\theta), \overline{v(\theta)}) + \mu(u_2(\theta), \overline{v(\theta)}),$$

and moreover,

$$(u(\theta), \overline{e^{in\theta}}) = a_n.$$

Define

$$(u(\theta), \overline{v(\theta - \psi)}) = (u(\theta + \psi), \overline{v(\theta)}) = \sum_n a_n \bar{b}_n e^{in\psi}$$

to be the *convolution* of the two functions $u(\theta)$, $v(\theta)$.

The most interesting example is the *Dirac function*

$$\delta(\theta) = \sum_n e^{in\theta}.$$

It gives us

$$(u(\theta), \overline{\delta(\theta - \psi)}) = \sum_n a_n e^{in\psi} = u(\psi).$$

We shall sometimes refer to $\delta(\theta - \psi)$ as $\delta_\psi(\theta)$.

The derivative of the generalized function $u(\theta)$ is defined to be

$$i \sum_n n a_n e^{in\theta},$$

and is denoted by $u'(\theta)$. Clearly

$$(u'(\theta), \overline{v(\theta)}) = i \sum_n n a_n \bar{b}_n = -\sum_n a_n \overline{(inb_n)}$$

$$= -(u(\theta), \overline{v'(\theta)}).$$

Therefore, we have

$$(u(\theta), \overline{\delta'(\theta - \psi)}) = -(u'(\theta), \overline{\delta(\theta - \psi)}) = -u'(\psi),$$

and

$$(u(\theta), \delta^{(v)}(\theta - \psi)) = (-1)^v u^{(v)}(\psi).$$

This is, however, too broad a definition for generalized functions, for it is impossible to derive from it many useful conclusions. We now introduce two special classes of generalized functions.

If the coefficient a_n satisfies

$$\max\left(\varlimsup_{n \to \infty} |a_n|^{1/n}, \varlimsup_{n \to \infty} |a_{-n}|^{1/n} \right) \leqslant 1,$$

then the corresponding generalized function is called a *generalized function of class H* or *of type H*.

It is clear that the generalized functions of class H form a linear set, and moreover their derivatives are still generalized functions of class H. The convolution of two generalized functions of type H is also a generalized function of type H.

If there exists an integer p, such that $a_n = O(|n|^p)$, then the corresponding generalized function is called a generalized function of type S; this is the generalized function of L. Schwarz.[2]

A generalized function of type S is obviously necessarily a generalized function of type H; furthermore, generalized functions of type S also form linear sets and they are closed under differentiation and convolution.

Remark. Within the class of generalized functions of type H, we defined two operations, "addition" and "convolution". If we look upon "convolution" as "multiplication", then the class of all generalized functions of type H forms a ring.

To be even more explicit, define

$$u(\theta) \pm v(\theta) = \sum_n (a_n \pm b_n) e^{in\theta},$$

$$u(\theta) \bigcirc v(\theta) = (u(\psi), v(\theta - \psi)) = \sum_n a_n b_n e^{in\theta}.$$

It is not difficult to prove directly the commutative, associative and distributive laws. Furthermore, this ring has the unit element

$$\delta(\theta) = \sum_n e^{in\theta};$$

in addition,

$$\lambda u(\theta) = (u(\theta), \lambda \delta(\psi - \theta)).$$

$e^{in\theta}$ $(n = 0, \pm 1, \pm 2, \ldots)$ is an idempotent, that is, $e^{in\theta} \bigcirc e^{in\theta} = e^{in\theta}$. These idempotents are mutually orthogonal, that is, $e^{im\theta} \bigcirc e^{in\theta} = 0$ (if $m \neq n$), and their sum is the unit element $\delta(\theta)$.

This property also holds for generalized functions of type S.

---

[2] i.e., in $\mathbb{R}^2$ (translator's note).

8.2 Duality

If two classes of generalized functions, T and $\overset{\circ}{T}$ satisfy the following three conditions, then they are said to be *dual* to each other:

   (i) $(u, \bar{v})$ is convergent for all $u \in T$ and for all $v \in \overset{\circ}{T}$.
  (ii) if $(u, \bar{v})$ is convergent for all $u \in T$, then $v \in \overset{\circ}{T}$.
 (iii) if $(u, \bar{v})$ is convergent for all $v \in \overset{\circ}{T}$, then $u \in T$.

EXAMPLE 1. The class K formed by all generalized functions and the class $\overset{\circ}{K}$ formed by all finite Fourier series are dual to each other.

EXAMPLE 2. Let $p > 1$ and $p' = p/(p - 1)$, then L^p and $L^{p'}$ are in duality. Here, L^p denotes the class of functions $u(\theta)$ which satisfy

$$\sum_n |a_n|^p < \infty.$$

EXAMPLE 3. If we replace the above requirement of the convergence of $\sum_n a_n \bar{b}_n$ by its (Cesàro) $(C, 1)$ summability, then we get in addition the following dual classes: (a) B and L^1 (where B denotes the class of Fourier series of bounded functions), (b) C and St form a pair of dual classes (where C denotes the Fourier series of continuous functions and St the Fourier-Stieltjes series).
    In summing, we have the following relation:

$$C \subset L^\infty = B \subset L^{p'} \subset L^2 \subset L^p \subset L \subset St.$$

It is easy to show that a class which coincides with its dual must be L^2.

Theorem 1. *Let $\varphi(n)$ represent an increasing positive function, so that as $n \to \infty$, $\varphi(n) \to \infty$. Moreover, assume that for any $\delta > 0$, the series*

$$\sum_{n=1}^{\infty} \frac{1}{(\varphi(n))^\delta}$$

always converges. If T denotes the class of all generating functions which satisfies

$$\log |a_n| = o(\log \varphi(|n|)),$$

and $\overset{\circ}{T}$ denotes the class of generating functions which satisfies

$$\log \varphi(|n|) = O\left(\log \frac{1}{|b_n|}\right),$$

then T and $\overset{\circ}{T}$ are in duality.

PROOF. (i) From the definition, we know that given any $\varepsilon > 0$, for n sufficiently large,

$$|a_n| \leq (\varphi(|n|))^\varepsilon,$$

moreover, there exists a number $c > 0$ such that

$$|b_n| \leqslant \frac{1}{(\varphi(|n|))^c}.$$

Hence $\sum_n a_n \bar{b}_n$ converges.

(ii) Assume $v \notin \mathring{T}$, then there exists a sequence n_v such that

$$\lim_{v \to \infty} \frac{\log \varphi(|n_v|)}{\log \dfrac{1}{|b_{n_v}|}} = \infty.$$

Choose $a_{n_v} = 1/b_{n_v}$ and the rest of the a_n to be 0; then the generalized function $u(\theta)$ so defined belongs to T while $\sum_n a_n \bar{b}_n$ diverges.

(iii) Assume $u \notin T$, then there exists a sequence n_v such that

$$\log |a_{n_v}| \geqslant c \log \varphi(|n_v|), \qquad c > 0.$$

Choose $b_{n_v} = 1/a_{n_v}$ and the rest of the b_n to be 0; then again the generalized function $v(\theta)$ thus defined belongs to $\mathring{T}$ while $\sum_n a_n \bar{b}_n$ diverges. $\quad\square$

If in theorem 1 we choose $\varphi(n) = e^n$, then the class T coincides with class H, for from

$$\log |a_n| = o(|n|),$$

we know that

$$\varlimsup_{n \to \infty} |a_n|^{1/n} \leqslant 1.$$

Now the relation

$$|n| = O\left(\log \frac{1}{|b_n|}\right),$$

is equivalent to

$$\varlimsup_{|n| \to \infty} |b_n|^{1/|n|} < 1.$$

Hence:

Theorem 2. *The dual class $\mathring{H}$ of the class H consists of the generalized functions which satisfy*

$$\max\left(\varlimsup_{n \to \infty} |b_n|^{1/n}, \ \varlimsup_{n \to \infty} |b_{-n}|^{1/n}\right) < 1.$$

By choosing $\varphi(n) = e^{n^p}$ ($p > 1$) we denote the resulting class T by G_p, and from theorem 1 we deduce:

Theorem 3. *If G_p is the class formed by the generalized functions which satisfy*

$$\varlimsup_{|n| \to \infty} |a_n|^{|n|^{-p}} \leqslant 1$$

then its dual, $\mathring{G}_p$, is formed by the generalized functions which satisfy

$$\varlimsup_{|n| \to \infty} |b_n|^{|n|^{-p}} < 1.$$

The class G_p was introduced by Gelfand-Shilov.

With a proof similar to that of theorem 1, we obtain:

Theorem 4. *Let $\varphi(n)$ represent an unbounded increasing positive function of n. Furthermore, assume that there exists a positive number λ such that*

$$\sum_{n=1}^{\infty} \frac{1}{(\psi(n))^{\lambda}}$$

converges. Let T represent the class of generalized functions which satisfies

$$\log|a_n| = O(\log \psi(|n|)),$$

and $\overset{\circ}{T}$ the class of generalized functions which satisfies

$$\log \psi(|n|) = o\left(\log \frac{1}{|b_n|}\right),$$

then T and $\overset{\circ}{T}$ are in duality.

By choosing $\psi(n) = n$ in the preceding theorem, the class T becomes class S, and we thereby have:

Theorem 5. *The dual class $\overset{\circ}{S}$ of the class S consists of the generalized functions which satisfy the condition that, for any $q > 0$.*

$$b_n = O\left(\frac{1}{|n|^q}\right).$$

In theorem 4 we again choose $\psi(n) = e^n$. We denote the resulting class T by I, and from theorem 4 we have:

Theorem 6. *The dual class $\overset{\circ}{I}$ of the class I is the class of all generalized functions which satisfy*

$$\lim_{n \to \infty} |b_n|^{-|n|^{-1}} = \infty$$

Remark 1. These classes can be further subdivided in a more refined way; for example, if in theorem 4 we change the condition to read

$$\overline{\lim_{n \to \infty}} \frac{\log|a_n|}{\log \psi(|n|)} \leqslant \rho,$$

then there is no difficulty whatsoever in finding its dual class.

Remark 2. It is very simple to prove that the class $\overset{\circ}{H}$ (or the class $\overset{\circ}{S}$) is also a linear set and that it is closed under differentiation and convolution. However, although it is a ring, it does not contain a unit element. $\overset{\circ}{H}$ is an ideal in H. Let H^* denote any ideal within H. If the generalized functions contained within H^* are all ordinary functions, that is to say if the formal Fourier series of these generalized functions are all convergent, then H^* is called a *function-ideal*. It is easily seen that $\overset{\circ}{H}$ is the maximal function-ideal in H.

8.3 Significance of the Generalized Functions of Type H

Corresponding to a generalized function $u(\theta)$ of type H, there exists a function

$$u(r, \theta) = \sum_n a_n e^{in\theta} r^{|n|}, \qquad 0 \leqslant r < 1, \qquad 0 \leqslant \theta < 2\pi,$$

this being a harmonic function on the unit disc.

Therefore a generalized function of type H can be looked upon as the boundary value of a harmonic function on the unit disc.

By the same reasoning, a generalized function $v(\theta)$ of type $\overset{\circ}{H}$ (i.e. a function in the usual sense) corresponds to a harmonic function on a larger concentric disc.

Another obvious fact is that corresponding to a generalized function $v(\theta)$ of type $\overset{\circ}{I}$, there exists a function

$$\sum_n b_n e^{in\theta} r^{|n|}$$

which is everywhere harmonic, and this type of function is also called an *entire harmonic function*. Therefore the generalized functions of type $\overset{\circ}{I}$ are the numerical values of the entire harmonic functions on the unit circle.

The generalized function

$$\delta_\psi(\theta)$$

belongs to H, but does not belong to $\overset{\circ}{H}$. The entire harmonic function corresponding to this generalized function is

$$\sum_n e^{in(\theta - \psi)} r^{|n|} = \frac{1 - r^2}{1 - 2r \cos(\theta - \psi) + r^2},$$

which is the well known Poisson kernel, $P(r, \theta)$.

Let $f(\theta)$ be a continuous (or integrable) function with Fourier series b_n, then for any $u(\theta) \in H$ we have

$$\frac{1}{2\pi} \int_0^{2\pi} u(r, \theta) \overline{f(\theta)} \, d\theta = \frac{1}{2\pi} \int_0^{2\pi} \sum_n a_n e^{in\theta} r^{|n|} \overline{f(\theta)} \, d\theta$$

$$= \sum_n a_n r^{|n|} \frac{1}{2\pi} \int_0^{2\pi} \overline{f(\theta)} e^{in\theta} \, d\theta$$

$$= \sum_n a_n \bar{b}_n r^{|n|}.$$

If $(u(\theta), \overline{f(\theta)})$ converges, then from Abel's theorem we know that

$$(u(\theta), \overline{f(\theta)}) = \lim_{r \to 1} \frac{1}{2\pi} \int_0^{2\pi} u(r, \theta)\overline{f(\theta)}\, d\theta.$$

In the more general case we have that for two generalized functions $u(\theta)$ and $v(\theta)$ of H, and for $r < 1, r' < 1$,

$$\frac{1}{2\pi} \int_0^{2\pi} u(r, \theta)\overline{v(r', \theta)}\, d\theta = \sum_n a_n \overline{b_n}(rr')^{|n|}.$$

If $(u, \overline{v})$ exists, then

$$(u, \overline{v}) = \lim_{r \to 1} \frac{1}{2\pi} \int_0^{2\pi} u(r, \theta)\overline{v(r, \theta)}\, d\theta.$$

Furthermore, for $u \in H$ and $v \in \overset{\circ}{H}$, when there exists a $\delta > 0$ such that $0 \leqslant r' < 1 + \delta$, then $v(r', \theta)$ is harmonic. By choosing $r = 1/(1 + \frac{1}{2}\delta)$ and $r' = 1 + \frac{1}{2}\delta$, we then know that

$$\frac{1}{2\pi} \int_0^{2\pi} u(r, \theta)\overline{v(r', \theta)}\, d\theta = \sum_n a_n \overline{b_n} = (u(\theta), \overline{v(\theta)}).$$

8.4 Significance of the Generalized Functions of Type S

Any Fourier series a_n of a continuous function $u(\theta)$ satisfies

$$|a_n| \leqslant \frac{1}{2\pi} \int_0^{2\pi} |f(\theta)|\, d\theta = O(1).$$

Now let there be a generalized function of type S; the Fourier series $a_n^{(p)}$ of its pth order derivative then satisfies

$$a_n^{(p)} = O(|n|^p).$$

Conversely, if

$$a_n = O(|n|^p),$$

then $u(\theta) - a_0$ is the $(p + 2)$th order derivative of the generalized function

$$\sum_n' \frac{a_n}{(in)^{p+2}} e^{in\theta},$$

Moreover this series converges uniformly to a continuous function. Thus, the class of generalized functions of type S actually consists of those generalized functions which are the derivatives of all orders of continuous functions.

We can similarly prove that the class of generalized functions of type $\overset{\circ}{S}$ is the set of infinitely differentiable functions.

From the results of the preceding paragraph we already know the following: if $u \in S$ and $v \in \overset{\circ}{S}$, then

$$(u, \bar{v}) = \lim_{r \to 1} \frac{1}{2\pi} \int_0^{2\pi} u(r, \theta)\overline{v(\theta)}\, d\theta.$$

If $u(\theta) - a_0$ is the pth order derivative of the continuous function $w(\theta)$, then by integration by parts we have

$$\begin{aligned}
(u, \bar{v}) &= a_0\bar{b}_0 + \lim_{r \to 1} \frac{1}{2\pi} \int_0^{2\pi} (u(r, \theta) - a_0)\overline{v(\theta)}\, d\theta \\
&= a_0\bar{b}_0 + \lim_{r \to 1} \frac{(-1)^p}{2\pi} \int_0^{2\pi} w(r, \theta)\overline{v^{(p)}(\theta)}\, d\theta \\
&= a_0\bar{b}_0 + \frac{(-1)^p}{2\pi} \int_0^{2\pi} w(\theta)\overline{v^{(p)}(\theta)}\, d\theta,
\end{aligned}$$

and therefore this is something we could have derived by the ordinary operations.

8.5 Annihilating Sets

Definition. An open interval $a < \theta < b$ of the unit circle is called an *annihilating interval* of a given generalized function of type H, $u(\theta)$, if in any closed subinterval of $a < \theta < b$, we have the uniform convergence

$$\lim_{r \to 1} u(r, \theta) = 0.$$

A point θ_0 is called a *point of support* of $u(\theta)$ if no annihilating interval of $u(\theta)$ contains θ_0. The union of all annihilating intervals of $u(\theta)$ is called the *annihilating set* of $u(\theta)$. This is an open set; we call its complement the *support* of $u(\theta)$. Obviously every point of the support of $u(\theta)$ is a point of support of $u(\theta)$.

EXAMPLE. $\delta_\psi(\theta)$ is a generalized function which has $\theta = \psi$ as its unique point of support. The reason for this is as follows: when $\theta \neq \psi$ and when $r \to 1$,

$$P(r, \theta) = \frac{1 - r^2}{1 + 2r \cos(\theta - \psi) + r^2}$$

tends to 0.

Theorem 1. *The generalized function of type H which has only one point of support, $\theta = \psi$, can be expressed as*

$$u(\theta) = \sum_{v=0}^{\infty} C_v \overline{D^{(v)}[\delta_\psi(\theta)]},$$

where $D^{(v)}[\delta_\psi(\theta)]$ is a linear combination of $\delta_\psi^0(\theta) = \delta_\psi(\theta),\ \delta'_\psi(\theta),\ \dots,\ \delta_\psi^{(v)}(\theta)$ and $\delta_\psi^{(v)}(\theta)$ is the vth order derivative of $\delta_\psi(\theta)$. More explicitly, for any $v \in \mathring{H}$, we have

$$(u(\theta), \overline{v(\theta)}) = \sum_{v=0}^{\infty} c_v \overline{D^{(v)}(v(\psi))},$$

where

$$D^{(v)}(v(\psi)) = \frac{1}{v!} \left[\frac{d^v v(\psi + \sin^{-1} x)}{dx^v} \right]_{x=0},$$

and both series are convergent.

Theorem 2. *A generalized function of type S which has only one point of support, $\theta = \psi$, can be expressed as*

$$u(\theta) = \sum_{v=0}^{l} c_v \overline{\delta^{(v)}(\theta)},$$

where l is a nonnegative integer. More explicitly, we have that for any $v \in \mathring{S}$,

$$(u(\theta), \overline{v(\theta)}) = \sum_{v=0}^{\infty} c_v \overline{v^{(v)}(\theta)}.$$

The proofs of the two theorems are as follows:

Without any loss in generality we assume $\psi = 0$, and that the unit circle is the interval $[-\pi, \pi]$ with endpoints identified.

For any given $\varepsilon > 0$, when $r \to 1$, the function tends to 0 uniformly for $|\theta| > \varepsilon$. Hence

$$\lim_{r \to 1} \int_{-\pi}^{\pi} u(r, \theta) \overline{v(\theta)} \, d\theta = \lim_{r \to 1} \int_{-\varepsilon}^{\varepsilon} u(r, \theta) \overline{v(\theta)} \, d\theta.$$

When ε is sufficiently small, for $|\theta| \leqslant \varepsilon$ the function $v(\theta)$ has an expansion

$$v(\theta) = \sum_{v=0}^{l} D^{(v)}(v(0)) \sin^v \theta + R(\theta),$$

where $R(\theta) = O(\varepsilon^{l+1})$, and

$$D^{(v)}(v(0)) = \frac{1}{v!} \left(\frac{d^v v(\sin^{-1} x)}{dx^v} \right) \Big|_{x=0}.$$

Under the assumption of theorem 1, this expansion remains valid with $l = \infty$, and when $|\theta| \leqslant \varepsilon$, it converges uniformly.

Therefore, since

$$
\begin{aligned}
C_\nu &= \lim_{r \to 1} \int_{-\varepsilon}^{\varepsilon} (\sin\theta)^\nu u(r,\theta)\,d\theta = \lim_{r \to 1} \int_{-\pi}^{\pi} (\sin\theta)^\nu u(r,\theta)\,d\theta \\
&= \lim_{r \to 1} \int_{-\pi}^{\pi} \left(\frac{e^{i\theta} - e^{-i\theta}}{2i} \right)^\nu u(r,\theta)\,d\theta \\
&= \lim_{r \to 1} \frac{1}{(2i)^\nu} \sum_{t=0}^{\nu} \binom{\nu}{t} \int_{-\pi}^{\pi} e^{it\theta} e^{-i(\nu-t)\theta} u(r,\theta)\,d\theta \\
&= \frac{1}{(2i)^\nu} \sum_{t=0}^{\nu} \binom{\nu}{t} a_{\nu-2t},
\end{aligned}
$$

we have proved theorem 1 (and we have in addition an expression for c_ν).$\Box$

Now to prove theorem 2, we let

$$
R^*(\theta) = \begin{cases} R(\theta) & \text{for } |\theta| < \varepsilon, \\ 0 & \text{otherwise.} \end{cases}
$$

Then

$$
\begin{aligned}
\int_{-\varepsilon}^{\varepsilon} R(\theta) u(r,\theta)\,d\theta &= \int_{-\pi}^{\pi} R^*(\theta) u(r,\theta)\,d\theta \\
&= \int_{-\pi}^{\pi} R^*(\theta)(u(r,\theta) - a_0)\,d\theta + a_0 \int_{-\varepsilon}^{\varepsilon} R(\theta)\,d\theta.
\end{aligned}
$$

The last term converges to 0 as $\varepsilon \to 0$. Now let $w(\theta)$ be a continuous function whose lth derivative equals $u(r,\theta) - a_0$. Then we get

$$
\begin{aligned}
\lim_{r \to 1} &\left| \int_{-\pi}^{\pi} R^*(\theta)(u(r,\theta) - a_0)\,d\theta \right| \\
&= \lim_{r \to 1} \left| \int_{-\pi}^{\pi} R^{*(l)}(\theta) w(r,\theta)\,d\theta \right| \leq \int_{-\pi}^{\pi} \left| R^{*(l)}(\theta) w(\theta) \right|\,d\theta = O(\varepsilon),
\end{aligned}
$$

thus proving theorem 2. $\Box$

8.6 Generalized Functions of Other Types

Definition. Let $\rho > 0$. If

$$
\overline{\lim_{n \to \infty}} \frac{\log |a_n|}{n \log n} < \frac{1}{\rho}, \qquad \overline{\lim_{n \to \infty}} \frac{\log |a_{-n}|}{n \log n} < \frac{1}{\rho},
$$

then $u(\theta)$ is called *a generalized function of class J_ρ* or *type J_ρ*.

It is clear that the generalized functions of class J_ρ form a linear set and that their derivatives are also generalized functions. However we must note the fact that the convolution of two generalized functions of class J_ρ is not necessarily a generalized function.

Theorem 1. *The generalized functions of the dual class $\mathring{J}_\rho$ satisfy the conditions*

$$\lim_{n\to\infty} \frac{\log \frac{1}{|b_n|}}{n \log n} \geqslant \frac{1}{\rho}, \qquad \lim_{n\to\infty} \frac{\log \frac{1}{|b_{-n}|}}{n \log n} \geqslant \frac{1}{\rho}.$$

PROOF. (1) We know from the definition of J_ρ, that for a $\delta > 0$, when there exists an $n_0(\delta)$ such that $n \geqslant n_0(\delta)$, then

$$\frac{\log |a_n|}{n \log n} < \frac{1}{\rho} - \delta,$$

that is,

$$|a_n| < n^{(1/\rho - \delta)n}.$$

On the other hand, from the definition of $\mathring{J}_\rho$, when n is sufficiently large,

$$|b_n| < n^{-(1/\rho - \delta/2)n},$$

and therefore $\sum_n a_n b_n$ converges.

(2) Next assume that

$$\overline{\lim_{n\to\infty}} \frac{\log |a_n|}{n \log n} \geqslant \frac{1}{\rho},$$

then there exists a sequence n_v such that

$$\overline{\lim_{v\to\infty}} \frac{\log |a_{n_v}|}{n_v \log n_v} = \frac{1}{\sigma} \geqslant \frac{1}{\rho}.$$

Choosing $b_{n_v} = 1/a_{n_v}$ and all of the rest of the b_n to be 0, the result is that $v(\theta) \in \mathring{J}_\rho$ and $\Sigma a_n b_n$ is divergent.

(3) Now assume that

$$\lim_{n\to\infty} \frac{\log \frac{1}{|b_n|}}{n \log n} < \frac{1}{\rho},$$

then there exists a sequence n_v such that

$$\lim_{n_v\to\infty} \frac{\log \frac{1}{|b_{n_v}|}}{n_v \log n_v} = \frac{1}{\tau} < \frac{1}{\rho}.$$

Choosing $a_{n_v} = 1/b_{n_v}$ and the rest of the a_n to be 0, we obtain a generalized function $u(\theta) \in J_\rho$, and $\Sigma a_n b_n$ diverges.

Corresponding to a generalized function $u(\theta)$ of class J_ρ, we introduce a function

$$u_\rho(r, \theta) = \sum_{m=0}^{\infty} P_m r^m \sum_{|n| \leqslant m} a_n e^{in\theta},$$

where
$$p_m = 1/(m!)^{1/\rho}.$$

Now since there exists a $\delta > 0$ such that

$$\left| p_m \sum_{|n| \leqslant m} a_n e^{in\theta} \right| \leqslant \frac{\sum_{|n| \leqslant m} |a_n|}{(m!)^{1/\rho}} = O\left(\frac{m \cdot m^{(1/\rho - \delta)m}}{m^{m/\rho}} \right) = O(m^{-\delta m/2}),$$

we know that on any compact subset of the plane, $u_\rho(r, \theta)$ is a uniformly (absolutely) converging series.

Theorem 2. *For any $u \in J_\rho$ and any $v \in \mathring{J}_\rho$,*

$$(u, \bar{v}) = \lim_{r \to \infty} \frac{1}{2\pi} \int_0^{2\pi} u_\rho(r, \theta) v(\theta) \, d\theta / J_\rho(r),$$

where

$$J_\rho(r) = \sum_{m=1}^{\infty} \frac{r^m}{(m!)^{1/\rho}}.$$

PROOF. It is very easily seen that

$$\frac{1}{2\pi} \int_0^{2\pi} u(r, \theta) \overline{v(\theta)} \, d\theta = \sum_l \sum_{n=0}^{\infty} p_m r^m \sum_{|n| \leqslant m} a_n \bar{b}_l \frac{1}{2\pi} \int_0^{2\pi} e^{i(n-l)\theta} \, d\theta$$

$$= \sum_{m=0}^{\infty} p_m r^m \sum_{|n| \leqslant m} a_n \bar{b}_n.$$

The theorem then follows from the generalized Borel summability theorem (see theorem 3 of the appendix to this chapter). $\qquad\square$

Similarly, we may define annihilating sets, that is, for a closed subinterval of the open interval $a < \theta < b$, when $r \to \infty$,

$$u_\rho(r, \theta)/J_\rho(r)$$

uniformly tends to 0. The union of all such open intervals is called the *annihilating set* of the generating function $u(\theta)$, and the points of the complement of the annihilating set are called *points of support*.

Remark. From our knowledge of the theory of entire functions, we know that $\mathring{J}_\rho$ is the set of all entire harmonic functions of order $\leqslant \rho$, and that the method of summation in J_ρ is just that of generalized Borel sums.

We have previously mentioned that the convolution of two generalized functions of type J_ρ is not necessarily a generalized function of type J_ρ. If we consider, however, the set J of generalized functions which satisfies

$$\log |a_n| = O(|n| \log |n|),$$

then the generalized functions of type J have the following three properties: (a) they form a linear set, (b) they are closed under differentiation and (c) they are closed under convolution. Furthermore, its dual $\mathring{J}$ is comprised of all of the entire functions of order zero.

8.7 Continuation

It is possible to further broaden our discussion. Let

$$Q(r) = \sum_{n=0}^{\infty} q_n r^n, \qquad q_n \geqslant 0$$

be a power series which converges for all r.

Let I_Q denote the set of all generalized functions which satisfy the conditions

$$\overline{\lim_{n \to \infty}} \frac{\log(|a_n|q_n)}{n \log n} < 0, \qquad \overline{\lim_{n \to \infty}} \frac{\log(|a_{-n}|q_n)}{n \log n} < 0.$$

We can prove by the same method:

Theorem 1. *The dual class $\mathring{I}_Q$ of the class I_Q consists of the generalized functions which satisfy*

$$\lim_{n \to \infty} \frac{\log\left(\dfrac{q_n}{|b_n|}\right)}{n \log n} \geqslant 0, \qquad \lim_{n \to \infty} \frac{\log\left(\dfrac{q_n}{|b_{-n}|}\right)}{n \log n} \geqslant 0.$$

It is also simple to prove that, for $u \in I_Q$ and $v \in \mathring{I}_Q$,

$$(u, \bar{v}) = \lim_{r \to \infty} \frac{1}{2\pi} \int_0^{2\pi} u_Q(r, \theta) \overline{v(\theta)} \, d\theta / Q(r),$$

where

$$u_Q(r, \theta) = \sum_{m=0}^{\infty} q_m r^m \sum_{|n| \leqslant m} a_n e^{in\theta}.$$

Given any kind of generalized function, we can always construct suitable q_n such that

$$\overline{\lim_{n \to \infty}} \frac{\log(|a_n|q_n)}{n \log n} < 0, \qquad \overline{\lim_{n \to \infty}} \frac{\log(|a_{-n}|q_n)}{n \log n} < 0.$$

In other words, no matter how complicated a generalized function, we always have a method of dealing with it.

8.8 Limits

The method of defining limits in the generalized functions of class T is as follows: we first assume that T consists entirely of ordinary functions.

Let $u_\nu(\theta)$ ($\nu = 1, 2, \ldots$) be a sequence of functions in T. $u(\theta) \in T$ is called the *limit* of this sequence if: for any function $v(\theta)$ in the dual class $\mathring{T}$, we have

$$\lim_{\nu \to \infty} (u_\nu, \bar{v}) = (u, \bar{v}).$$

As for the generalized functions in the dual class $\mathring{T}$ of T, they are generally functions in the ordinary sense. Depending on the situation, we will define limits in $\mathring{T}$ as limits of functions in the ordinary sense.

(i) Suppose $v_\nu(r, \theta)$ $(\nu = 1, 2, \ldots)$ is a sequence of functions which are harmonic on an open region containing the unit disc, and which are in addition uniformly convergent on any compact subset in this region. Then their limit function $v(r, \theta)$ is obviously harmonic in this region. We call $v(\theta)$ the limit of $v_\nu(\theta)$ and we denote it by $v_\nu(\theta) \to v(\theta)(\mathring{H})$.

For any $u(\theta) \in H$, we choose $\delta\,(>0)$ sufficiently small, such that the common domain of definition of each of $v_\nu(r, \theta)$ contains the closed disc of radius $r' = 1 + \frac{1}{2}\delta$. From §3 we know that

$$(u(\theta), v_\nu(\theta)) = \frac{1}{2\pi} \int_0^{2\pi} u(r, \theta)\overline{v_\nu(r', \theta)}\, d\theta,$$

$$r = \frac{1}{1 + \frac{1}{2}\delta}.$$

By hypothesis $v_\nu(r', \theta)$ converges uniformly in the disc of radius r'; we therefore have

$$\lim_{\nu \to \infty} (u, \overline{v}_\nu) = \frac{1}{2\pi} \int_0^{2\pi} u(r, \theta)\overline{v(r', \theta)}\, d\theta = (u, \overline{v}).$$

(ii) Suppose $v_\nu(r, \theta)$ $(\nu = 1, 2, \ldots)$ is a sequence of functions which are harmonic on the unit disc, infinitely differentiable on the unit circle and furthermore, for each $p \geqslant 0$, the sequence

$$\frac{\partial^p}{\partial \theta^p} v_\nu(r, \theta)$$

converges uniformly on the closed unit disc. Then it is not difficult to deduce that its limit function $v(r, \theta)$ also is harmonic on the unit disc and is infinitely differentiable on the unit circle. We define this to be $v_\nu(\theta) \to v(\theta)(\mathring{S})$.

Take a generalized function

$$u(\theta) = \sum_n a_n e^{in\theta},$$

which belongs to class S, and write

$$v_\nu(\theta) = \sum_n b_n^{(\nu)} e^{in\theta}, \qquad v(\theta) = \sum_n b_n e^{in\theta}.$$

Since $v_\nu(\theta)$ converges uniformly on the unit circle, we have

$$\lim_{\nu \to \infty} b_0^{(\nu)} = \lim_{\nu \to \infty} \frac{1}{2\pi} \int_0^{2\pi} v_\nu(\theta)\, d\theta = \frac{1}{2\pi} \int_0^{2\pi} v(\theta)\, d\theta = b_0.$$

Since we can assume from §4 that $u(\theta)$ is the pth derivative of the continuous function $w(\theta)$, we have

$$(u, \overline{v}_\nu) = a_0 b_0^{(\nu)} + \frac{(-1)^p}{2\pi} \int_0^{2\pi} w(\theta)\overline{v_\nu^{(p)}(\theta)}\, d\theta.$$

Since we have assumed that $v_v^{(p)}(\theta)$ also converges uniformly to $v^{(p)}(\theta)$, therefore

$$\lim_{v \to \infty} (u, \bar{v}_v) = a_0 b_0 + \frac{(-1)^p}{2\pi} \int_0^{2\pi} w(\theta) \lim_{v \to \infty} \overline{v_v^{(p)}(\theta)} \, d\theta = (u, \bar{v}).$$

In other words, if

$$v_v(\theta) \to v(\theta)(\mathring{S}),$$

then we always have[3]

$$\lim_{v \to \infty} (u, \bar{v}_v) = (u, \bar{v}).$$

(iii) Suppose $v_v(r, \theta)$ $(v = 1, 2, \ldots)$ is a sequence of functions harmonic on the entire plane, and that on any compact subset, this sequence converges uniformly to a function $v(r, \theta)$. We define

$$v_v(\theta) \to v(\theta)(\mathring{I}),$$

so that, from the proof of (i) we necessarily have

$$\lim_{v \to \infty} (u, \bar{v}_v) = (u, \bar{v}).$$

(iv) Suppose $v_v(r, \theta)$ $(v = 1, 2, \ldots)$ is a sequence of entire harmonic functions of order zero which on any compact subset of the plane converges uniformly to a function $v(r, \theta)$, also of zero order. Then we may define $v_v(\theta) \to v(\theta)(\mathring{J})$, and obtain similarly,

$$\lim_{v \to \infty} (u, \bar{v}_v) = (u, \bar{v}).$$

We see from the several examples above, that corresponding to the dual class $\mathring{T}$ of any class of generalized functions T, we can always use similar methods to appropriately introduce the limit concept within $\mathring{T}$, so that the following relations hold:

$$\lim_{v \to \infty} (u, \bar{v}_v) = (u, v).$$

We leave it to the reader to enumerate all the other cases.

8.9 Addenda

(A) From the standpoint of the theory of complex functions of one variable, the reason we introduce these generalized functions is very natural. We have

---

[3] It is not difficult to prove that the limit defined here of the class $\mathring{S}$ and that defined by L. Schwarz are the same. It is worth noting that when we defined the generalized function of type S of Schwarz, we did not at all assume continuity of this arbitrary function, and in so doing, we have proved that its continuity was inevitable and was not required in the hypothesis of the definition.

Class	Definition of the class
$\overset{\circ}{K}$	finite sum (harmonic polynomials)
$\overset{\circ}{G}_p$	entire harmonic functions of zero order and type p
$\overset{\circ}{J}$	zero order entire harmonic functions
$\overset{\circ}{I}$	all entire harmonic functions
$\overset{\circ}{H}$	those harmonic functions whose domains of regularity include the closed unit disc in the interior

Moreover, we have the relation

$$\overset{\circ}{K} \subset \overset{\circ}{J} \subset \overset{\circ}{I} \subset \overset{\circ}{H} \subset H \subset I \subset J \subset K.$$

Making use of the order of entire functions, we can insert J_ρ between $\overset{\circ}{J}$ and $\overset{\circ}{I}$.

Making use of the properties of boundary values of harmonic functions we can insert other classes between $\overset{\circ}{H}$ and H, such as those listed in example 3 of §2.

On the other hand, from the standpoint of the summation theory of divergent series, our discussion also has its significance.

(B) From theorems 1 and 4 of §2 we know that classification by means of the order of growth of the coefficients is also very natural and exhaustive. The principle classes are as follows:

Class	The order of $\log|a_n|$
S	$O(\log n)$
H	$o(n)$
I	$O(n)$
J	$O(n \log n)$
G_p	$o(n^p),\ p > 1$

(C) From the basic theorem on conformal mappings, the definition of the class H is not limited to the unit disc. We can consider in any plane a simply connected domain which has more than one boundary point, in particular the upper half plane with the real axis as boundary.

On the other hand, from the similarity between Fourier series and Fourier integrals, we can directly consider the "formal Fourier integral"

$$u(x) = \frac{1}{\sqrt{2\pi}} \int_{-\infty}^{\infty} a(t)e^{itx}\, dt.$$

If $\log|a(t)| = o(t)$, then $u(x)$ is said to belong to the class H, and if $\log|a(t)| = O(\log|t|)$, then $u(x)$ is said to belong to class S. The corresponding harmonic function of $u(x)$ is

$$u(x, y) = \frac{1}{\sqrt{2\pi}} \int_{-\infty}^{\infty} a(t)e^{itx - |t|y}\, dt, \qquad y > 0.$$

(D) To generalize even more, we begin by extending theorems 1 and 4 of §2: let $\mathbb{R}^n$ be the n-dimensional real Euclidean space, let $t = (t_1, \ldots, t_n)$ be the points in $\mathbb{R}^n$, and let

$$\tau = \sqrt{t_1^2 + \cdots + t_n^2}.$$

We have the following result which is similar to theorem 1.

Let $\varphi(\tau)$ be a positive increasing function of one variable, $\tau (\geqslant 0)$, and for any $\delta > 0$ assume the integral

$$\int_0^\infty (\varphi(\tau))^{-\delta} \tau^{-1} \, d\tau \tag{1}$$

is convergent. Let A denote the functions $a(t)$ which satisfy the following conditions: (i) in any bounded region $a(t)$ is square integrable, and (ii) outside a set of measure zero, when τ is sufficiently large,

$$\log |a(t)| = o(\log \varphi(\tau)). \tag{2}$$

Now let B denote the functions $b(t)$ which satisfy the following conditions: (i) in any bounded region $b(t)$ is square integrable, and (ii) outside a set of measure zero, when τ is sufficiently large,

$$\log \varphi(\tau) = O\left(\log \frac{1}{|b(t)|} \right). \tag{3}$$

Then, A and B share the following three properties: (i) if $a(t) \in A$ and $b(t) \in B$, then

$$\int_{-\infty}^\infty \cdots \int a(t)\overline{b(t)} \, dt_1 \cdots dt_n < \infty, \tag{4}$$

(ii) if for any $b(t) \in B$ expression (4) converges, then $a(t) \in A$, (iii) if for any $a(t) \in A$ expression (4) converges, then $b(t) \in B$.

The proof of this is quite clear. Now we have the result which is similar to theorem 4.

Let $\overset{\circ}{K}$ be the generalized function formed by

$$v(x) = \frac{1}{(\sqrt{2\pi})^n} \int_{-\infty}^\infty \cdots \int b(t) e^{-itx'} \, dt_1 \cdots dt_n$$

where $tx' = t_1 x_1 + \cdots + t_n x_n$, and the integral is absolutely convergent. Consider the integral

$$\int_{-\infty}^\infty \cdots \int a(t)\overline{b(t)} \, dt$$

as a linear functional acting on A. Then it defines a generalized function $u(x)$ such that

$$(u(x), \overline{v(x)}) = \int_{-\infty}^\infty \cdots \int a(t)\overline{b(t)} \, dt_1 \cdots dt_n,$$

and $u(x)$ may be expressed by the formal Fourier integral

$$u(x) = \frac{1}{(\sqrt{2\pi})^n} \int_{-\infty}^{\infty} \cdots \int a(t) e^{-itx'} \, dt_1 \cdots dt_n.$$

Such a $u(x)$ can be explicitly given by the following methods:
  (a) if for any $\varepsilon > 0$, $a(t) = O(e^{\varepsilon|t|})$, then use

$$u(x, y) = \frac{1}{(\sqrt{2\pi})^n} \int_{-\infty}^{\infty} \cdots \int a(t) e^{-itx' - |t|y'} \, dt_1 \cdots dt_n,$$

where $|t| = (|t_1|, \ldots, |t_n|)$, and

$$(u(x), \overline{v(x)}) = \lim_{y \to 0} \int_{-\infty}^{\infty} \cdots \int u(x, y) \overline{v(x)} \, dx.$$

  (b) Under other conditions we introduce

$$u(x, r) = \int_0^\infty \frac{r^v \, dv}{\varphi(v)} \int_{\tau < v} a(t) e^{-itx'} \, dt \Big/ \int_0^\infty \frac{r^v \, dv}{\varphi(v)},$$

and

$$(u(x), \overline{v(x)}) = \lim_{r \to \infty} \int_{-\infty}^{\infty} \cdots \int u(x, r) \overline{v(x)} \, dx_1 \cdots dx_n.$$

This method appears to have some marked advantages over the method of L. Schwarz.
  (c) If we use any partial differential equation of elliptic type in place of the Laplace equation, there corresponds to its solutions within the domain a theory of generalized functions which is, naturally, even more general. Such examples can be found in the author's work on the harmonic functions of classical domains and the Fourier analysis of characteristic manifolds, etc.

Appendix: Summability

Theorem 1. *Let $q_v(r)$ be a sequence of functions which are defined on an interval with r_0 as its right endpoint, and suppose it has the following properties:*

  (i) $q_v(r) \geqslant 0$,
  (ii) $\sum_{v=0}^{\infty} q_v(r) = 1$, $\quad r < r_0$,
  (iii) $\lim_{r \to r_0} q_v(r) = 0$, $\quad$ for any v,

then from the fact that $s_n \to s$, we can deduce that

$$\lim_{r \to r_0} \sum_{v=0}^{\infty} q_v(r) s_v = s.$$

PROOF. We may assume, without any loss in generality, that $s = 0$. Then for any given $\varepsilon > 0$, there exists an M such that when $v \geqslant M$, $|s_v| < \varepsilon$. We can further assume that for all v, $|s_v| \leqslant B$. We then have

$$\left| \sum_{v=0}^{\infty} q_v(r) s_v \right| \leqslant B \sum_{v=0}^{M} q_v(r) + \varepsilon \sum_{v=M+1}^{\infty} q_v(r)$$

$$\leqslant B \sum_{v=0}^{M} q_v(r) + \varepsilon,$$

so that when $r \to r_0$, the right side tends to ε. $\qquad\square$

Let us now consider two important examples:

(a) Choose $r_0 = 1$, $q_v(r) = (1 - r)r^v$, then conditions (i), (ii) and (iii) are all satisfied, and so when $r \to 1$,

$$\sum_{v=0}^{\infty} (1 - r)r^v s_v = \sum_{v=0}^{\infty} (s_v - s_{v-1})r^v \to \lim_{v \to \infty} s_v.$$

and this is just:

Theorem 2 (Abel). *If*

$$\sum_{v=0}^{\infty} a_v$$

converges to s, then when $r \to 1$,

$$\sum_{v=0}^{\infty} a_v r^v$$

also tends to s.

(b) Suppose

$$\sum_{v=0}^{\infty} p_v r^v, \qquad p_v \geqslant 0$$

is an everywhere convergent power series. Let

$$q_v(r) = p_v r^v \left/ \sum_{v=0}^{\infty} p_v r^v \right.,$$

then (i) and (ii) clearly hold. Now since

$$\lim_{r \to \infty} q_v(r) = \lim_{r \to \infty} \frac{v p_v r^{v-1}}{\sum_{v=1}^{\infty} v p_v r^{v-1}} = \cdots = 0,$$

we have:

Theorem 3 (Borel). *If*

$$\sum_{v=0}^{\infty} p_v r^v, \qquad p_v \geqslant 0,$$

is an everywhere convergent power series, and if $s_\nu \to s$, then

$$\lim_{r \to \infty} \frac{\sum\limits_{\nu=0}^{\infty} p_\nu r^\nu s_\nu}{\sum\limits_{\nu=0}^{\infty} p_\nu r^\nu} = s.$$

Theorem 1 has an integral analogue.

Theorem 4. *Let $q(r, \theta)$ be a function defined in the intervals $0 \leqslant \theta \leqslant 2\pi$ and $0 \leqslant r < 1$ which has the following properties:*

(i) $q(r, \theta) \geqslant 0$,

(ii) $\int_0^{2\pi} q(r, \theta)\, d\theta = 1$,

and for any $\varepsilon > 0$,

$$\lim_{r \to 1} \int_{|\theta| > \varepsilon} q(r, \theta)\, d\theta = 0.$$

Suppose when $\theta \to \pm 0$, $f(\theta) \to s$, then

$$\lim_{r \to 1} \int_0^{2\pi} q(r, \theta) f(\theta)\, d\theta = s.$$

(we are assuming here that $f(\theta)$ is a bounded function).

Index